Experimental
Malignant Hyperthermia

Charles H. Williams
Editor

Experimental Malignant Hyperthermia

With 39 Figures

Springer-Verlag
New York Berlin Heidelberg
London Paris Tokyo

Charles H. Williams, Ph.D.
Director of Surgery Research
Texas Tech University Health Sciences Center
Regional Academic Health Center at El Paso
Department of Surgery
School of Medicine
El Paso, Texas 79905, USA

Library of Congress Cataloging-in-Publication Data
Experimental malignant hyperthermia.
  Includes bibliographies and index.
  1. Malignant hyperthermia.   2. Physiology,
Experimental.   I. Williams, Charles H.
(Charles Herbert), 1935–   . [DNLM:
1. Anesthesia–adverse effects.   2. Disease Models,
Animal.   3. Malignant Hyperthermia–etiology.
WO 245 E96]
RD82.7.M3E97   1987   617′.96   87–26443
ISBN-13:978-1-4612-8327-0

Typeset by David Seham Associates, Metuchen, New Jersey.

9 8 7 6 5 4 3 2 1

ISBN-13:978-1-4612-8327-0   e-ISBN-13:978-1-4612-3738-9
DOI:10.1007/978-1-4612-3738-9

Preface

The major portion of this book is the result of a continuous series of experimental investigations with an inbred colony of malignant hyperthermia (MH) susceptible pigs that was initially identified and established at the University of Wisconsin Arlington Swine Farm in July 1970. Five of the original bred females and seven weaning pigs were moved to the University of Missouri Sinclair Comparative Medicine Research Farm in July 1973. Nine, seventh and eighth generation MH susceptible pigs were moved to the Clint Experimental Medicine Research Farm in October 1982, where the MH susceptible swine colony is currently maintained.

The ready availability of highly susceptible MH swine has been a key component in our experimental investigations of malignant hyperthermia during the past sixteen years. We have observed that inbreeding the MH-positive animals produces individuals that are stress susceptible and that will develop the MH syndrome from exercise, high ambient temperature, transport stress, or the psychological stress of being placed in a strange, dark, chamber, i.e., the Missouri Partitional Calorimeter.

Our experimental work has shown that the genetics of MH produces a spectrum of abnormal physiological responses ranging from normal to severe hypertension (over 400 Torr), normal to severe hypermetabolism (over $10 \times$ normal), normal to severe pulmonary hypertension, a variable range of responses to halothane anesthesia, and an abnormal myoneural junction with an increased voltage that lasts for a longer period of time when the nerve depolarizes the muscle.

Our studies of muscle relaxants suggest that nondepolarizing muscle relaxants are safe to use whereas depolarizing muscle relaxants are potent MH triggering agents. Our studies of general anesthetic agents indicate that halothane is a potent triggering agent but that other general anesthetics may be weaker and slower triggering agents.

Our investigations of MH at the cellular level suggest that a massive release of catecholamines (primarily norepinephrine) induces a rapid shift in metabolism toward accelerated heat production in muscle that rapidly leads to ATP depletion, lactic acidosis, glycogenolysis, and rapid muscle cell death in untreated animals. The massive catecholamine release undoubtedly affects calcium levels in muscle cells, probably via the alpha receptor through the phosphatidyl pathway to release inositol-triphosphate, a known calcium activator.

Studies in the Missouri Partitional Calorimeter clearly demonstrate that the MH genetic defect causes a high rate of heat production in the intact, untreated, animal which leads to high metabolic rates in vivo. The heat is produced in muscle tissue via a substrate or *futile cycle*. The futile cycle continues at a high rate postmortem which leads to rapid glycogenolysis, lactic acidosis, and heat buildup in certain muscles to produce pale, soft, exudative muscle. The precise, detailed, biochemical mechanism for the futile cycle is not readily apparent from the published studies or our own research although substrate cycling at the fructose-6-phosphate $\rightarrow$ fructose-1,6-bisphosphate level could be involved in an energy sink that would result in accelerated glycolysis. We can also postulate that a change in sodium ion and/or calcium ion permeability, which may be induced by norepinephrine hormonal activity on muscle cells to shift the metabolic set point towards heat production via nonshivering thermogenesis, would be a key control point for regulating muscle cell function. Our experimental observation that a 1.83-fold increase in motor unit potential voltage that lasts 1.12 times longer indicates that the motor nerve–muscle regulatory interaction is abnormal. Further studies at the cellular and subcellular levels are needed in order to elucidate the genetically induced regulatory and functional abnormality that causes malignant hyperthermia.

Chapters from other MH investigators have been included in this book in order to provide the reader with an excellent source and reference work for future experimental malignant hyperthermia investigations.

Charles H. Williams
El Paso, Texas
February, 1987

Acknowledgment

This book is the direct result of a Symposium on Hormonal, Physiologic, and Clinical Studies of Factors Affecting Heat Production During Malignant Hyperthermia held during the American Physiological Society Fall Meeting at Niagara Falls, New York, October 13–18, 1985.

We are especially grateful to the sponsors of the symposium who provided the funds to defray all costs of organizing and conducting the symposium. They were: American Physiology Society/Environmental, Thermal, and Exercise Physiology Section; Marion Laboratories, Inc; Boehringer Ingeheim Animal Health, Inc; Norwich Eaton Pharmaceuticals, Inc; Vital Signs, Inc; Astra Pharmaceutical Products, Inc; Burroughs Wellcome Co.; and The RGK Foundation.

I must also thank Dolores B. Williams, my wife, for her total support of this project from the time of initiation of the symposium through the final publication of this completed volume. Jim Hastings deserves special thanks for his art work for the symposium.

Contents

Preface v

1 Heat Production in Malignant Hyperthermia Susceptible Muscle 1
Charles H. Williams

2 Plasma Catecholamines During Malignant Hyperthermia 7
Charles H. Williams, Susan E. Dozier, Charles W. Gehrke, Klaus O. Gerhardt, and Joseph K. Wong

3 Measurement of Biogenic Amines by HPLC-ED: ECD As a
Diagnostic Tool for Malignant Hyperthermia 20
Klaus O. Gerhardt, Charles W. Gehrke, Charles H. Williams, and Joseph K. Wong

4 Hemodynamics in Malignant Hyperthermia Susceptible Pigs During
Malignant Hyperthermia 30
Charles H. Williams, Susan E. Dozier, and Mary Farias

5 Plasma Levels of T_4, T_3, and rT_3 During Malignant Hyperthermia .. 46
Ralph R. Anderson, Muftah A. Akasha, David A. Nixon, and Charles H. Williams

6 Malignant Hyperthermia and the Sarcoplasmic Reticulum Membrane:
A Review 59
Mariam A. Marvasti and Charles H. Williams

7 In Vitro Studies of Drugs Affecting Malignant Hyperthermia
Muscle 78
Wilfried K. Ilias, Charles H. Williams, Susan E. Dozier, and Robert T. Fulfer

8   The Role of the Horse in Studies Relative to Malignant
    Hyperthermia 91
    Charles E. Short and Nora S. Matthews

9   Horses and Ponies as Animal Models for Malignant Hyperthermia ... 100
    Susan V. Hildebrand

10 Malignant Hyperthermia in the Dog: Laboratory Investigations 118
    Peter H. Cribb

11 Laboratory Methods for Malignant Hyperthermia Diagnosis 121
    Jeffrey E. Fletcher and Henry Rosenberg

12 Malignant Hyperpyrexia: A Review 141
    Michael Denborough

13 Malignant Hyperthermia Pre- and Post-Dantrolene: A Survey of the
    Greater Kansas City Area from 1965 to 1985 147
    Mark G. Zukaitis, George P. Hoech, Jr., and John D. Robinson

14 The Role of the Sympathetic Nervous System in Patients Susceptible
    to Malignant Hyperthermia 155
    *J. Hilary Green, F. Richard Ellis, P. Jane Halsall, and
    Iain T. Campbell*

    Index 161

Contributors

Muftah A. Akasha, M.S., Research Assistant, Department of Dairy Science, University of Missouri Columbia, Columbia, Missouri, USA

Ralph R. Anderson, Ph.D., Professor, Department of Dairy Science, University of Missouri Columbia, Columbia, Missouri, USA

Iain T. Campbell, M.D., F.F.A.R.C.S., Senior Lecturer, University of Liverpool, Liverpool Royal Infirmary, Liverpool, England

Peter H. Cribb, B.Sc., M.R.C.V.S., Diplomate, A.C.V.A., Department of Veterinary Anesthesiology, Radiology, and Surgery, Western College of Veterinary Medicine, University of Saskatchewan, Saskatoon, Saskatchewan, Canada

Michael Denborough, M.D., Department of Medicine and Clinical Science, John Curtin School of Medical Research, Australian National University, Canberra, Canberra, Australia

Susan E. Dozier, B.S., Research Assistant, Department of Anesthesiology, Texas Tech University Health Sciences Center, Regional Academic Health Center, El Paso, Texas, USA

F. Richard Ellis, Ph.D., F.F.A.R.C.S., Reader in Anaesthesia, University Department of Anaesthesia, St. James University Hospital, Leeds, England

Mary Farias, M.T., Laboratory Supervisor, Departments of Anesthesiology and Biochemistry, Texas Tech University Health Sciences Center, Regional Academic Health Center, El Paso, Texas, USA

Jeffrey E. Fletcher, Ph.D., Assistant Professor, Department of Anesthesiology and Pharmacology, Hahnemann University, Philadelphia, Pennsylvania, USA

Robert T. Fulfer, M.D., Instructor, Department of Anesthesiology, Texas Tech University Health Sciences Center, Regional Academic Health Center, El Paso, Texas, USA

Charles W. Gehrke, Ph.D., Professor, Department of Biochemistry, Manager, Experiment Station Chemical Laboratories, University of Missouri, Columbia, Missouri, USA

Klaus O. Gerhardt, Dr. Rer. Nat., Senior Research Chemist, Department of Biochemistry, Experiment Station Chemical Laboratories, University of Missouri, Columbia, Missouri, USA

J. Hilary Green, Ph.D., Lecturer, Department of Human Movement, University of Western Australia, Perth, Australia

P. Jane Halsall, M.B., Ch.B., Research Clinical Assistant, University Department of Anaesthesia, St. James University Hospital, Leeds, England

Susan V. Hildebrand, D.V.M., Associate Professor, Department of Surgery, School of Veterinary Medicine, University of California, Davis, California, USA

George P. Hoech, Jr., M.D., Clinical Professor, Department of Anesthesiology, University of Missouri Medical School at Kansas City, Missouri, Vice President, Medical Staff, Research Medical Center, Kansas City, Missouri, USA

Wilfried K. Ilias, M.D., Klinik f.Anaesthesie u.allgem.Intensivmedizin, d.Univ. Wien, Austria, and Visiting Associate Professor (1984) Department of Anesthesiology, Texas Tech University Health Sciences Center, Regional Academic Health Center, El Paso, Texas, USA

Mariam A. Marvasti, M.D., Research Associate, Departments of Anesthesiology and Biochemistry, Texas Tech University Health Sciences Center, Regional Academic Health Center, El Paso, Texas, USA

Nora S. Matthews, D.V.M., Section of Anesthesiology, Department of Clinical Sciences, New York State College of Veterinary Medicine, Cornell University, Ithaca, New York, USA

David A. Nixon, B.S., Research Assistant, Department of Dairy Science, University of Missouri Columbia, Columbia, Missouri, USA

John D. Robinson, M.D., Staff Anesthesiologist, Shawnee Mission Medical Center, Shawnee Mission, Kansas, and Department of Anesthesiology, University of Kansas Medical Center, Kansas City, Kansas, USA

Henry Rosenberg, M.D., Professor and Chairman, Department of Anesthesiology, Hahnemann University, Philadelphia, Pennsylvania, USA

Charles E. Short, D.V.M., M.S., D.A.C.V.A. Section of Anesthesiology, Department of Clinical Sciences, New York State College of Veterinary Medicine, Cornell University, Ithaca, New York, USA

Charles H. Williams, Ph.D., Associate Professor, Departments of Surgery and Biochemistry, Director of Surgery Research, Texas Tech University Health Sciences Center, Regional Academic Health Center, El Paso, Texas, USA

Joseph K. Wong, M.S., Department of Biochemistry, Experiment Station Chemical Laboratories, University of Missouri, Columbia, Missouri, USA

Mark G. Zukaitis, M.D., Assistant Clinical Professor, Department of Anesthesiology, University of Missouri School of Medicine at Kansas City, Missouri and University of Kansas Medical Center, Kansas City, Kansas. Chairman, Department of Anesthesia, Research Medical Center, 2316 E Meyer Blvd, Kansas City, Missouri, USA

CHAPTER 1

Heat Production in Malignant Hyperthermia Susceptible Muscle

Charles H. Williams

Introduction

Malignant hyperthermia (MH) in man and the porcine stress syndrome (PSS) in pigs appear to be the same genetic defect, since the same drugs, for example, such general anesthetics as halothane and such depolarizing muscle relaxants as succinylcholine, will trigger the onset of MH. The malignant hyperthermia susceptible (MHS) pig has been used by our laboratory for 16 years as an animal model of the human MH syndrome.[1-7]

There are several key points that must be considered when developing a hypothesis to explain the generation of heat during MH: (1) The heat is produced in muscle, as shown by direct measurement of muscle and blood temperature,[2,6,7] and may be produced for up to 14 min before muscle rigor develops. (2) The muscles that are the most severely affected are the white type muscles, which use primarily glycogen as an energy source.[7] In the MHS pig these muscles are the longissimus dorsi, the gluteus medius, and the biceps femoris. (3) The heat generating mechanism leads to high basal metabolic rates in vivo[3,6,7] and accelerated glycogenolysis postmortem in the affected muscles to produce pale, soft, exudative muscle.[7] (4) Mitochondria isolated from MHS pig muscles do not have abnormal functional characteristics in vitro, that is, uncoupling that would explain the increased metabolic rate observed in vivo. (5) The heat-generating mechanism, energy sink, must continue to operate under anaerobic conditions at a high rate and to draw on reserve glycogen and creatine phosphate stores to provide adenosine triphosphate (ATP) for postmortem cellular activities. The rapid glycolysis observed by Kastenschmidt et al.[8] was a key observation in documenting this point. Our data suggesting that the energy sink could be a substrate cycle,[2] or futile cycle, were a step forward in elucidating the mechanism of heat generation. (6) The cardiovascular system is severely compromised in vivo, with extremely high heart rates and very high blood pressures.[7] The MHS pig can develop MH spontaneously from a number of environmental stress factors.[4,7,9] Malignant hyperthermia is not a pharmacogenetic disease but rather a genetic disease that can be triggered also by potent gaseous anesthetics, such as halothane, and by depolarizing muscle relaxants, such as succinylcholine.[4,7] (8) After 16 years of research on the MHS sarcoplasmic reticulum by various laboratories, there is scant evidence that the sarcoplasmic reticulum is involved in causing the MH syndrome.[10] (9) Prevention of plasmalemma depolarization by nondepolarizing muscle relaxants will prevent the development of MH.[11] (10) Stimulation of the alpha receptor with alpha agonists will cause MH in susceptible pigs,[12] whereas blockade of the alpha receptor with alpha antagonists will prevent the most lethal components of MH, which are vasoconstriction and muscle rigor.[13] (11) There is a massive, early release and accumulation of norepinephrine and epinephrine in the bloodstream (a 100- to 300-fold increase), which plays a key role in driving the MH syndrome to the lethal stages.[11-19] (12) Calcium channel blockade with diltiazem will abort the MH syndrome.[20] (13) Calcium ion levels in myoplasm reflect the increased metabolic activity of MHS muscle, but calcium ion per se does not have any intrinsic properties that can cause MH. The basic and fundamental problem with calcium ion appears to be the failure of MHS muscle to regulate the calcium ion activity properly. The increased acidity of MHS muscle[21] would play a big role in promoting increased calcium ion activity by the simple process of acid buffering in muscle cells. If other ionic activities were measured, for example, K^+, Na^+, Mg^{2+}, we would expect to see large fluctuations in those ionic levels also. Measuring calcium ion activity levels does not provide an absolute indication of calcium ion flux or turnover rates. (14) The MH genetic defect can be inherited at different dosage levels, thereby producing a broad range of pathophysiologic responses in susceptible pigs. The gene dosage in most human MH patients can be assumed to be heterozygous, since the normal marriage pattern is an outcross. Therefore, the MH gene in humans can be expected to produce pathophysiologic symptoms ranging from vasoconstriction in the hands and feet,[22] droopy eyelids,[23] muscle atrophy,[24] unexplained fevers,[25] labile hypertension,[26] increased drug sensitivity,[27] all the way up to a lethal reaction in the operating room.

Discussion

Figure 1.1 illustrates the important points about membrane functional characteristics in MHS people and provides a framework for further experimental investigations and pharmacologic studies. The acetylcholine receptor appears to be involved in MH, since physical exercise, depolarization with succinylcholine, and our studies of spontaneous myoneural junctional activity indicate that the motor unit potentials are elevated 1.83 times and last 1.12 times longer than normal.[28] The alpha receptor is involved as the primary site of neurohormonal stimulation by norepinephrine to activate the phosphatidyl inositol phosphate (PIP_2) system, a secondary

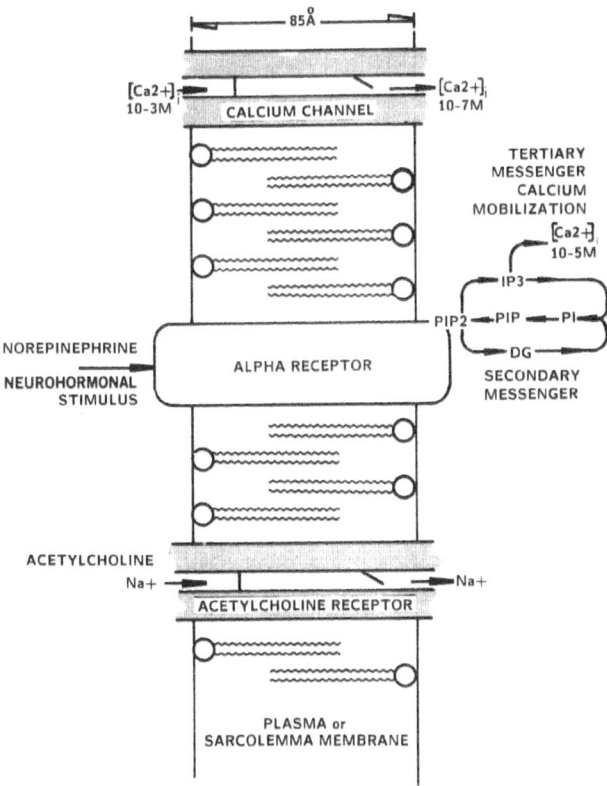

FIGURE 1.1. Schematic diagram of the interrelated events at the cell membrane that enable MH to develop.

messenger, inside the cell. The subsequent release of inositol triphosphate (IP$_3$) would act to mobilize calcium, a tertiary messenger, which could result in a series of calcium-activated cellular changes. Since the integrity of PIP$_2$ in the cellular membrane may also play a role in calcium gating,[29,30] the flow of calcium ion through the membrane via the calcium channel may be increased markedly when the PIP$_2$ system is activated or the cell membrane is depolarized. Thus, the concentration of ionized calcium inside the cell could be increased by two entirely different but additive mechanisms that would result in a net increase of calcium ion activity. The increased calcium ion activity could then result in increased calcium ion flux at various metabolic steps and result in increased metabolic rate via vectorial ion transport.[31]

Norepinephrine may induce increased thermogenesis via futile cycles, that is, metabolic pathways that result in net ATP hydrolysis. Further investigations are needed on the regulation of substrate or futile cycles.

These interrelationships are supported by the following studies. Calcium channel blockers, for example, diltiazem, will abort the MH syndrome in vivo[20] and is protective to MH muscle strips in vitro.[32] Phenylephrine, an alpha agonist, will induce MH in susceptible pigs in vivo.[12] Very high levels of norepinephrine (100- to 300-fold) have been measured quantitatively before and during MH in susceptible pigs.[11-18] Phentolamine, an alpha antagonist, will prevent the vasoconstriction and muscle rigor phases of MH in susceptible pigs.[13] Total myoneural blockade with metocurine[11] or pancuronium[33] will prevent the development of MH in susceptible pigs. Inspection of Figure 1.1 provides several areas for creative experimental investigations.

References

1. Williams CH, Galvez TL, Brucker RF, Popinigis J, Vail WJ (1971) Malignant hyperpyrexia in swine: A genetic disease of membrane function. Fed Proc 30:1208

2. Clark MG, Williams CH, Pfiefer WF, Bloxham DP, Holland PC, Taylor CA, Lardy HA (1973) Accelerated substrate cycling of fructose-6-phosphate in the muscle of malignant hyperthermic pigs. Nature 245:99–101

3. Williams CH, Houchins C, Shanklin MD (1975) Energy metabolism in pigs susceptible to the fulminant hyperthermia stress syndrome. Br Med J 3:411–413 and Br Med J 1:724 (1977) errata

4. Williams CH (1977) The development of an animal model for the fulminant hyperthermia-porcine stress syndrome, in Henschel EO (ed) *Malignant Hyperthermia: Current Concepts.* New York: Appleton-Century-Crofts, Chap 8, pp 117–140

5. Williams CH, Lasley JF (1977) The mode of inheritance of the fulminant hyperthermia stress syndrome in swine, in Henschel EO (ed) *Malignant Hyperthermia: Current Concepts.* New York: Appleton-Century-Crofts, Chap 9, pp 141–148

6. Williams CH, Shanklin MD, Houchin C (1977) Energy metabolism in fulminant hyperthermia-stress syndrome in swine, in Henschel EO (ed) *Malignant Hyperthermia: Current Concepts.* New York: Appleton-Century-Crofts, Chap 10, pp 149–156

7. Williams CH, Shanklin MD, Hedrick HB, Muhrer ME, Stubbs DH, Krause GF, Payne CG, Benedict JD, Hutcheson DP, Lasley JF (1978) The fulminant hyperthermia-stress syndrome: Genetic aspects, hemodynamic, and metabolic measurements in susceptible and normal pigs, in Aldrete JA, Britt BA (eds) *Proceedings 2nd International Symposium on Malignant Hyperthermia, Denver, April 1977.* New York: Grune & Stratton, pp-113–150

8. Kastenschmidt LL, Hoekstra WG, Briskey EJ (1968) Glycolytic intermediates and co-factors in "fast" and "slow-glycolyzing" muscles of pigs. J Food Sci 33:151–153

9. Topel DG, Bicknell EJ, Preston KS, Christain LL, Matsushima CY (1968) Porcine stress syndrome. Mod Vet Pract 49:40–60

10. Marvasti M, Williams CH (1987) The sarcoplasmic reticulum and malignant hyperthermia: A review, in Williams CH (ed) *Experimental Malignant Hyperthermia.* New York: SpringerVerlag pp 59–77

11. Hoech GP Jr, Roberts JT, Williams CH, Waldman SD, Simpson ST, Trim C, Brazile J (1980) Prevention of porcine malignant hyperthermia with metocurine, in Lomax P (ed) *Thermoregulatory Mechanisms and their Therapeutic Implications. 4th Interational Symposium on the Pharmacology of Thermoregulation, Oxford, 1979.* Basel: Karger, pp 137–141

12. Hall GM, Lucke JN, Lister D (1977) Porcine malignant hyperthermia V Fatal hyperthermia in the Pietrain pig associated with the infusion of alpha-adrenergic agonists. Anaesthesia 49:855–863

13. Williams CH, Stubbs DH (1977) Preliminary studies on the therapeutic efficacy of phentolamine in the fulminant hyperthermia-stress syndrome, in Lomax P (ed) *Drugs, Biogenic Amines, and Body Temperature. 3rd International Symposium on the Pharmacology of Thermoregulation. Banff, 1976.* Basel: Karger, pp 233–234

14. Davis TP, Gehrke CW Jr, Williams CH, Gehrke CW and Gerhardt KO (1982) Pre-Column derivatization and high performance liquid chromatography of biogenic amines in blood of normal and malignant hyperthermic pigs. J Chromatogr 228:113–122

15. Buzello, W, Williams CH, Chandra P, Watkins ML, Dozier SE (1985) Vecuronium and porcine malignant hyperthermia. Anesth Analg 64:515–519

16. Williams CH, Dozier SE, Buzello W, Gehrke CW, Wong JK, Gerhardt KO (1985) Plasma levels of norepinephrine and epinephrine during malignant hyperthermia in susceptible pigs. J Chromatogr Biomed Appl 344:71–80

17. Williams CH, Buzello W, Dozier SE, Joyner J (1984) Hemodynamics and oxygen use during malignant hyperthermia. Fed Proc 43(3):292

18. Williams CH, Buzello W, Dozier SE, Chandra P, Gehrke CW, Gerhardt KO, Wong JK (1985) Hemodynamics, cardiac function, oxygen use, and catecholamines during porcine malignant hyperthermia. Anesth Analg 64(2):1

19. Dozier SE, Williams CH, Ilias WK (1985) Norepinephrine potentiates contracture in malignant hyperthermia susceptible (MHS) porcine muscle. Fed Proc 44(5):1376

20. Williams CH, Dozier SE, Ilias WK, Fulfer RT, Zukaitis MG, Hoech GP Jr (1985) Treatment of malignant hyperthermia (MH) with diltiazem. Fed Proc 44(5):1638

21. Roberts JT, Burt T, Gyulai L, Screter F, Allen PA, Ryan JF, Chance B (1985) Delayed recovery of intracellular skeletal muscle pH measured by phosphorus-31 nuclear magnetic resonance in malignant hyperthermic swine with partial to full recovery of arterial pH following treatment with sodium dantrolene. Anesthesiology 63(3A):A272

22. Wingard DW, Gatz EE (1978) Some observations on stress susceptible patients, in Aldrete JA, Britt BA (eds) *Proceedings 2nd International Symposium on Malignant Hyperthermia, Denver, April 1977.* New York: Grune & Stratton, pp 363–372

23. Britt BA, Kalow W (1968) Hyperrigidity and hyperthermia associated with anesthesia. Ann NY Acad Sci 151:947–948

24. Denborough MA, Ebeling P, King JO, Zapf P (1970) Myopathy and malignant hyperpyrexia. Lancet 1:1138–1140

25. Wilson RD, Dent TE, Traber DL, McCoy NR, Allen CR (1967) Malignant hyperpyrexia with anesthesia. JAMA 202(3):183–186

26. Williams CH, Stubbs DH, Payne CG, Benedict JD (1976) Role of hypertension in fulminant hyperthermia-stress syndrome. Br Med J 1:628

27. Merchandani H, Reich LE (1985) Fatal MH as a result of ingestion of tranylcypromine combined with white wine and cheese. J Forensic Sci 30:217–220

28. Steiss JE, Bowen JM, Williams CH (1981) Electromyographic evaluation of malignant hyperthermia-susceptible pigs. Am J Vet Res 42(12):2061–2064

29. Agranoff BW (1986) Inositol trisphosphate and related metabolism. Fed Proc 45(11):2627–2628

30. Putney JW, Aub DL, Taylor CW, Merritt JE (1986) Formation and biological action of inositol 1,4,5-trisphosphate. Fed Proc 45(11):2634–2638

31. Harold FM (1986) *The Vital Force: A Study of Bioenergetics.* New York: WH Freeman & Co, pp 303–358

32. Ilias WK, Williams CH, Fulfer RT, Dozier SE (1985) Diltiazem inhibits halothane-induced contractions in malignant hyperthermia-susceptible muscles in vitro. Br J Anaesth 57:994–996

33. Jones DE, Ryan JF, Taylor B, Lopez JR, Alamo L, Sreter FA, Allen PD (1985) Pancuronium in large doses protects susceptible swine from halothane-induced malignant hyperthermia. Anesthesiology 63(3A):A344

Note added in proof: The ability of cyclic Guanosine MonoPhosphate (cGMP) to open cation-selective channels in retinal rod outer segments[1] and to set up a cGMP cascade[2,3] for amplification of the visual signal provides a model for cGMP to interact at the muscle cell plasma membrane receptors to produce heat via a futile cycle transduced by G proteins[4] responding to norepinephrine hormonal stimulation.

1. Zimmerman AL, Yamanaka G, Eckstein F, Baylor DA, Stryer L (1985) Interaction of hydrolysis-resistant analogs of cyclic GMP with the phosphodiesterase and light-sensitive channel of retinal rod outer segments. Proc Natl Acad Sci 82:8813–8817

2. Stryer L (1986) Cyclic GMP cascade of vision. Ann Rev Neurosci 9:87–119

3. Stryer L (1987) The molecules of vision. Sci Amer 257(1):42–50

4. Gilman AG (1987) G Proteins: Transducers of receptor-generated signals. Ann Rev Biochem 56:615–649

CHAPTER 2

Plasma Catecholamines During Malignant Hyperthermia

Charles H. Williams, Susan E. Dozier,
Charles W. Gehrke, Klaus O. Gerhardt,
and Joseph K. Wong

Introduction

The development of an intense peripheral vasoconstriction as measured by the Duke's bleeding time by Muhrer et al.[1] and the decrease of radiation heat losses in the Missouri Partitional Calorimeter[2] suggested that the development of malignant hyperthermia (MH) was driven by a potent hormone response.[3] A hormone with the potential to accelerate the metabolic rate and to produce an intense peripheral vasoconstriction effect was norepinephrine.[3] We explored the possibility that thyroid hormones also could be involved in the acceleration of heat production.[4] In 1973, the only analytic method available for catecholamines was the trihydroxyindole procedure.[5] We were aware of the analytic deficiencies inherent in the trihydroxyindole procedure and elected to pursue the development of a fluorescent analytic method.[6-8] The fluorescent analytic method produced data showing that norepinephrine increased markedly as the MH syndrome developed.[8] We recently used high performance liquid chromatography with electrochemical detection (HPLC-EC) to quantitate the level of plasma norepinephrine and epinephrine in malignant hyperthermia–susceptible (MHS) pigs as the MH syndrome was developing.[9]

Methods and Materials

Eleven MHS pigs from our genetic strain of halothane-sensitive pigs were tested with halothane at 8 to 10 weeks of age to determine MH susceptibility.[10] The animals were fed a 16% protein swine ration, and those used for these experiments weighed 52 to 62 kg. The unpremedicated pigs were anesthetized by intraperitoneal injection of thiopental (22 mg/kg). An ear vein was cannulated, and 2 to 5 mg/kg increments of thiopental were injected as needed. The animal was intubated and mechanically ventilated with 66% nitrous oxide (NO_2) in oxygen (O_2) in a semiclosed circuit with 5 L/minute fresh gas flow. The efficiency of carbon dioxide (CO_2) absorption was monitored continuously by a Cavitron infrared absorption capnograph. A 7F Opticath® thermodilution catheter was advanced to the wedge position in the pulmonary artery via the right femoral vein and connected to an Oximetrix Shaw catheter oximeter (OS127OA). A 16G catheter and a 4F Opticath® connected to a second Oximetrix device were inserted into the right femoral artery and advanced into the abdominal aorta. The continuous monitoring included EKG, heart rate, arterial pres-

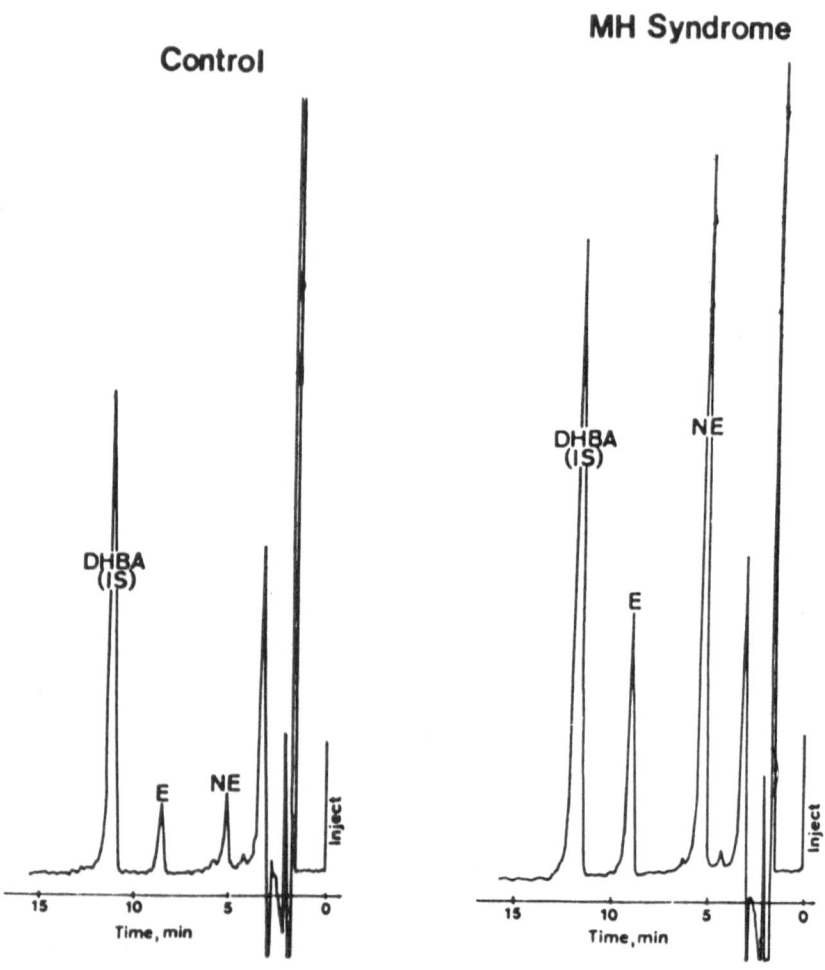

FIGURE 2.1. Reversed phase HPLC of norepinephrine and epinephrine in pig plasma by electrochemical detection. Sample: 30 μL = 0.1 ml plasma. Column:Supelcosil LC-18-DB, 5 μm, 150 mm × 4.6 mm. Mobil phase: citric acid, 0.025 M. Na,HPO$_4$: 5 × 10^{-5} M. EDTA: 35 mg octylsodium sulfate/L. MeOH: 3%. pH: 3.5. Flow rate: 1.2 ml/min. Detector: ECD LC-48B, 0.65 V Ag + /Ag/Cl reference electrode 2 nA full scale. Temperature: 31°C.

sure, pulmonary artery pressure, central venous pressure, arterial and mixed venous oxygen saturation, rectal and blood temperature, CO_2 concentration of the inhaled and exhaled gas, and evoked twitch tension. Cardiac output was determined every 10 minutes by an Edwards thermodilution cardiac output computer model 9520.

Further details on the physiologic monitoring and the results obtained are in our publication on Vecuronium and MH.[11] Five-milliliter (ml) blood samples were obtained from the external iliac vein, the pulmonary artery, and the femoral artery during the control phase of the experiment, at the first signal of tachycardia after adding 2% halothane to the rebreathing circuit, and at 5 to 10 minute intervals thereafter. Immediately after being drawn, each blood sample was mixed with 0.5 ml of freshly made ethylenediaminotetraacetate (EDTA)-metabisulfite solution, placed in an ice water bath, and processed as previously reported.[6–8]

The perchloric acid extract samples were cleaned up by alumina absorption prior to HPLC analysis for norepinephrine and epinephrine with Bioanalytical Systems Model LC-4B electrochemical detector at 0.65 V, 2 nA full scale after HPLC separation on a Supelcosil LC-18-DB, 150 mm × 4.6 mm column with an aqueous mobile phase of 0.025 M citric acid, 0.025 M Na_2HPO_4, 5×10^{-5} M EDTA, 35 mg octylsodium sulfate/L, and 3% MeOH, at pH 3.4. The flow rate was 1.2 ml/minute. An internal standard 3,4–dihydroxybenzylamine was added to the perchloric acid extract. Multiple blood samples were collected from 11 pigs susceptible to MH. Wilcoxon signed rank test and rank sum test statistical analyses were used for evaluating the data.

Results

The results in Figure 2.2 show that tachycardia increased significantly ($P < 0.05$) 5 minutes before any other symptoms of MH developed and at least 10 min before any increase of core temperature was observed. One pig in this group developed MH spontaneously from surgical stress, and the slight rise in core temperature was due to this one pig.[11] The increased heart rate response to halothane administration was not secondary to such factors as blood volume depletion, since intravenous fluids were continuously given and the central venous pressure was in the normal range (data not shown).

The results in Figure 2.3 show the changes in plasma levels of norepinephrine and epinephrine in Pig 1-4 during the control period, at the time tachycardia developed, and during the full course of the MH syndrome, including the development of muscle rigor. There was a significant elevation of the plasma level of norepinephrine, up to 50 ng/ml, at the first sign of tachycardia. The norepinephrine level further increased to a maximum of 85 ng/ml. This peak level coincided with the development of muscle rigor. The levels of norepinephrine dropped to 49 ng/ml, then

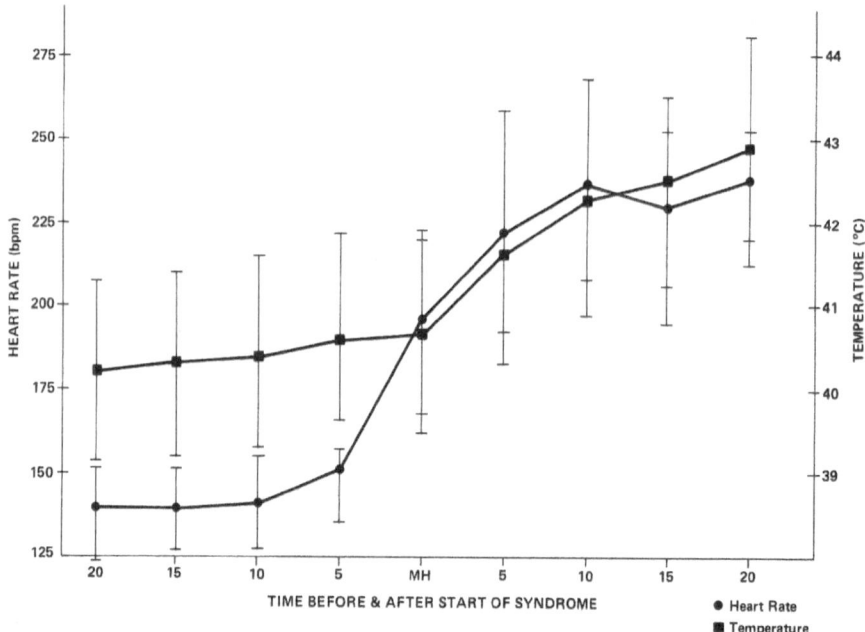

FIGURE 2.2. Temperature and heart rate during MH. Each point represents the mean value of data from six MHS pigs. Halothane was administered 10 minutes before MH started.

27 ng/ml, and 7 ng/ml as the cardiovascular collapse progressed. Epinephrine levels lagged behind and peaked later than norepinephrine levels. Figure 2.4 shows similar data from Pig 3-1. This pig developed a core temperature of 45°C without muscle rigor. The peak level of norepinephrine was 22 ng/ml of plasma. The pooled data in Table 2.1 show that all the pigs had low blood levels of norepinephrine and epinephrine during the control period of the experiment, that is, 0.35 to 0.64 ng/ml for norepinephrine and 0.21 to 0.59 ng/ml for epinephrine. At tachycardia, the norepinephrine level increased to 17 ng/ml and the epinephrine level increased to 4 ng/ml, giving a ratio of 4:1. The highest plasma level of norepinephrine we observed was 108 ng/ml.

At MH_1 and MH_2 the norepinephrine level was 39 ng/ml and 40 ng/ml, respectively. The epinephrine levels were 15 ng/ml and 24 ng/ml at the same times. The predominance of norepinephrine released early in the syndrome suggests that the peripheral adrenergic nerve secretion is activated first, with the adrenal gland secreting epinephrine on a slower time scale.

Table 2.1 also shows the data on four pigs that developed tachycardia that lasted for 1 hour without any other symptoms of MH. The increase

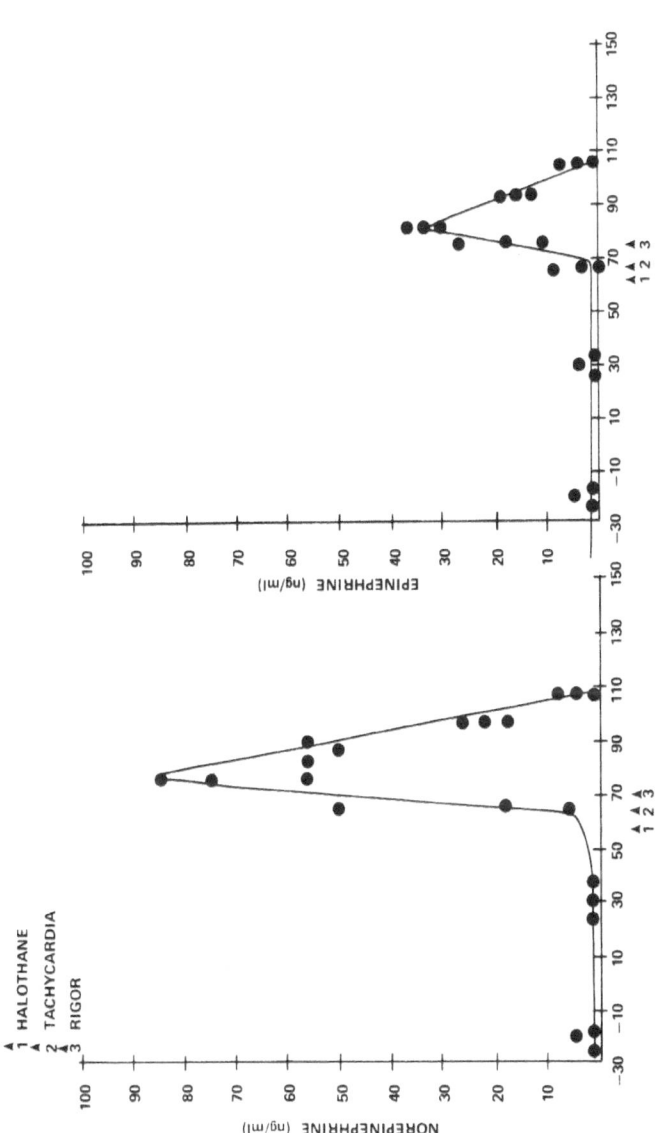

FIGURE 2.3. Plasma levels of norepinephrine and epinephrine in Pig 1–4 during halothane-induced MH. Each point represents the measured catecholamine values from independent analyses of serial samples of arterial, pulmonary artery, and venous plasma samples. This pig developed muscle rigor and MH. (From J. Chromatogr. 344:75–80, 1985. Reprinted with permission of Elsevier.)

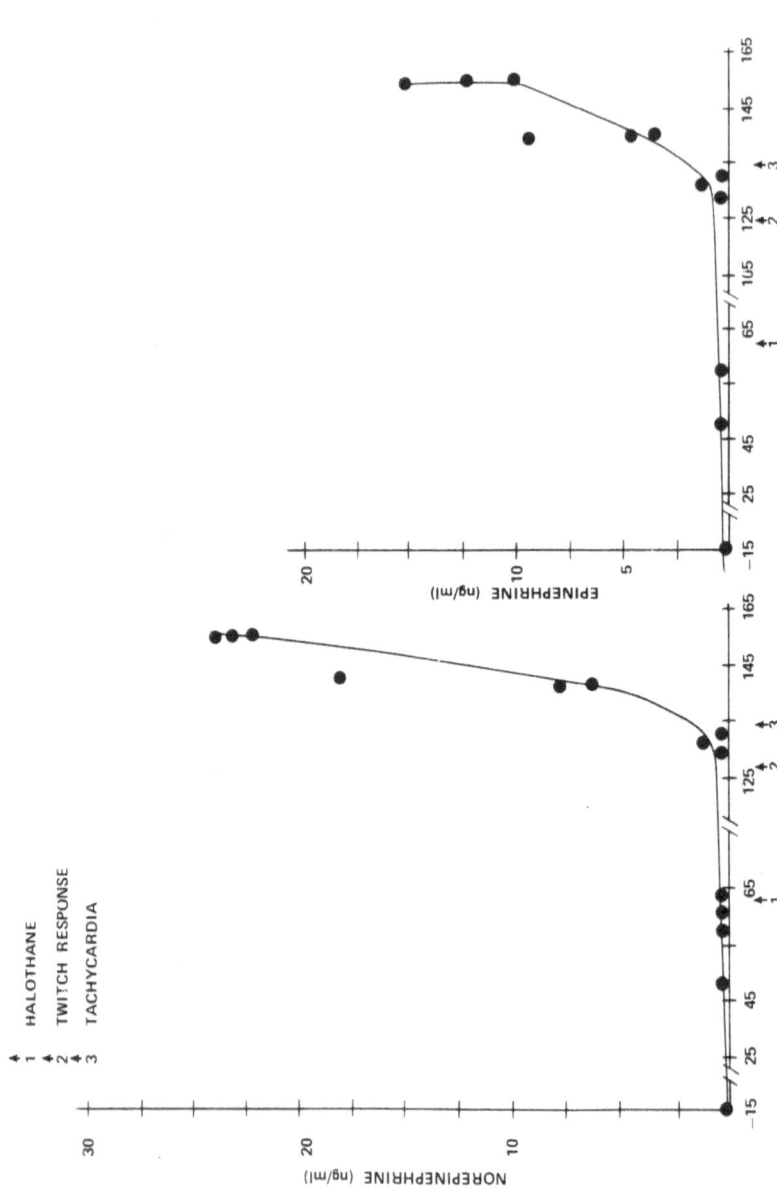

FIGURE 2.4. Plasma levels of norepinephrine and epinephrine in Pig 3–1 during halothane-induced MH. Each point represents the value obtained from independent analyses of serial samples of arterial, pulmonary artery, and venous plasma samples. This pig developed a core temperature of 45°C without muscle rigor or stiffness.

of norepinephrine in blood plasma was approximately 1.4 times ($P < 0.05$) the control values during the entire period of the tachycardia.

One pig was considered normal after being exposed to 2% halothane for 1 hour and injected with two separate doses of succinylcholine (1 mg/kg) in attempts to trigger the MH syndrome. There was a transient increase of norepinephrine in blood plasma (up to 1.44 ng/ml), accompanied by a short run of tachycardia that quickly reverted to normal heart rate, and then the norepinephrine in blood plasma decreased to control levels.

Discussion

We suggested in 1974 that the release of norepinephrine played a key role in initiating the development of the MH syndrome.[1,10] Because there was not a precise, specific, sensitive, and reproducible method for measuring plasma norepinephrine levels at that time, we developed and perfected a sensitive, selective, and reproducible HPLC fluorescent method.[6,7] Plasma norepinephrine data from MHS pigs showed a significant increase of norepinephrine as the syndrome developed.[8] With the advent of the electrochemical detector, enabling the analysis of norepinephrine and epinephrine simultaneously, we decided to reinvestigate the changes in plasma levels of norepinephrine and epinephrine during MH. The data from the recording system showed that tachycardia heralded the development of the MH syndrome 5 minutes sooner than any other parameter. Blood samples collected during the start of tachycardia showed significantly increased levels of norepinephrine. This initial release of norepinephrine occurs simultaneously with the increased heart rate.

Our data support the idea that norepinephrine mediates an alpha receptor process to cause vasoconstriction,[1,3,10] increased skeletal muscle contractility, heat generation via a nonshivering mechanism (i.e., futile or substrate cycling),[12] an increased metabolic rate,[2,13] and heat retention leading to hyperthermia. We have in vitro data on MH muscle strips demonstrating that norepinephrine potentiates the contractility of MH skeletal muscle.[14] The overall severity of the syndrome may be a reflection of the initial amount of norepinephrine released and subsequently accumulating as the syndrome progresses. In the liver, the stimulation of the alpha receptor causes a release of calcium ions, which act as secondary messengers intracellularly to initiate the changes in cellular metabolic processes.[15] A similar process could be occurring in muscle tissue. A further development of this rationale would entail the activation or opening of calcium channels by norepinephrine and epinephrine in skeletal muscle cell membranes, thereby allowing an increased calcium flux into the skeletal muscle cells, which would activate the ion translocation processes. If sufficient calcium enters the skeletal muscle cells, calcium activated muscle contraction (twitching) and/or rigor would result.

TABLE 2.1. Plasma norepinephrine and epinephrine levels during malignant hyperthermia[a]

MH syndrome pigs (n = 6)

	Control period				Halothane / MH period				
Time[b]	−40	−30	−20	−10	−5 Tachycardia	0 MH$_1$	+10 MH$_2$	+20 MH$_3$	+30 MH$_4$
Mean[c]	(15)	(14)	(7)	(7)	(16)	(15)	(5)	(3)	(3)
NE	0.58 ± 0.80	0.35 ± 0.05	0.36 ± 0.04	0.64 ± 0.41	17.49 ± 14.44	38.90 ± 30.26	40.27 ± 19.36	22.02 ± 5.1	3.29 ± 3.11
E	0.59 ± 0.88	0.21 ± 0.05	0.30 ± 0.0	0.35 ± 0.13	4.33 ± 3.48	14.98 ± 12.89	24.40 ± 11.93	17.81 ± 2.36	2.88 ± 2.04
Ratio NE:E	0.98:1	1.67:1	1.2:1	1.82:1	4.04:1	2.59:1	1.65:1	1.23:1	1.14:1

Wilcoxon signed rank test $P < 0.032$ Only halothane

Pigs (n = 4) Developing tachycardia (TC) control period — Tachycardia only

Time[b]	−30	−20	−10	TC	+10 TC$_2$	+20 TC$_3$	+30 TC$_4$	+40 TC$_5$
	(12)	(12)	(6)	(12)	(12)	(9)	(3)	(3)
NE	0.29 ± 0.01	0.30 ± 0.02	0.34 ± 0.08	0.44 ± 0.19	0.57 ± 0.24	0.37 ± 0.16	0.42 ± 0.12	0.36 ± 0.08
E	0.0 ± 0	0.0 ± 0	0.02 ± 0.06	0.14 ± 0.08	0.16 ± 0.06	0.08 ± 0.07	0.09 ± 0.08	0.0
Ratio NE:E				3.14:1	3.56:1	4.63:1	4.67:1	
Wilcoxon rank sum test				$P < 0.05$				

Normal pig ($n = 1$)

Time[b]	Control period		Halothane and succinylcholine
	-20	-10	
Mean	(3)	(3)	(3)
NE	0.29 ± 0	0.29 ± 0	0.29 ± 0
E	0 ± 0	0.29 ± 0	0.29 ± 0

E, epinephrine; NE, norepinephrine.

Numbers in parenthesis denote the total number of separate blood samples analyzed.

MH syndrome pig ($n = 7$) developed all the pathophysiologic indications of MH after exposure to 2% halothane. Tachycardia only pigs ($n = 4$) did not react to halothane but did develop a tachycardia after injection of succinylcholine (1 mg/kg). The tachycardia continued unabated for 1 hour and was treated then with Diltiazem (10 mg/pig). The normal pig ($n = 1$) did not react to halothane or to two injections of succinylcholine except to have a short burst of tachycardia that lasted 3 to 5 minutes.

Precision (from 15 independent analyses of pooled pig plasma)
  Norepinephrine 1.66% RS
  Epinephrine 2.06% RSD
Recovery (from 15 additions to pooled pig plasma)
  Norepinephrine 99.17%
  Epinephrine 93.33%

[a]Values are reported as nanograms of catecholamine per 1 ml of plasma.
[b]Time expressed in minutes, the minus sign indicates time prior to the MH syndrome.
[c]All data reported are mean ± SD.

Previous studies have shown that norepinephrine and epinephrine have a potentiating effect on induced muscle contractility in the isolated phrenic nerve–diaphragm preparation of the rat[16] and on skeletal muscle.[17] Investigations indicate that norepinephrine can open or activate calcium channels and thereby increase calcium flux across membranes.[18] The idea that norepinephrine can modulate calcium channel activity in stimulated cells[18] needs to be explored with experimental studies in MHS pigs. The level of norepinephrine that we have measured in the blood plasma of these pigs suggests that a maximal neurohormonal stimulation of all responsive cells would be effected. Our in vivo data showing high circulating levels of norepinephrine suggest that the intermittent release of norepinephrine that may occur during the susceptible animal's daily activities may hormonally induce the metabolic changes leading to the hypermetabolism and heat production that we previously reported[2,13] and may account for the increased amplitude and duration of the motor unit potentials we have observed.[19]

Our observations provide a rationale for the spontaneous development of the porcine stress syndrome (PSS), MH, triggered or induced by any type of physical activity or heat stress[20] and the development of pale, soft, and exudative muscle postmortem from stress–susceptible pigs.[21]

Analogous tachycardia has been recognized as a symptom of the MH syndrome in human patients for more than 20 years.[22–24] Based on our research, there is a greater need to recognize tachycardia as an early diagnostic sign of impending MH in human patients. Our experimental animal data suggest that detailed biogenic amine studies of human MH patients would be appropriate. A report of exercise physiology stress tests in MHS human patients implicates the sympathetic nervous system in the abnormal temperature responses observed.[25] Our data do not support other investigators' conclusions that the catecholamine response is secondary.[26,27] There are two reasons for the differences in the results: (1) We used halothane as the primary agent to trigger MH, with succinylcholine being used only if halothane did not produce the syndrome. Other groups used succinylcholine as the initial triggering agent.[26,27] However, we observed that the development of MH in response to an injection of succinylcholine was so rapid and compressed in time (1 to 3 minutes) that it was extremely difficult to determine the priority of the events. Therefore, we routinely used halothane as the triggering agent because it produced a slow motion series of pathophysiologic changes that could be resolved and adequately timed. (2) The trihydroxyindole method for measuring catecholamines has serious analytic deficiencies. Other reasons are listed in our Letter to the Editor.[28]

The preponderance of norepinephrine released in relation to epinephrine in MH suggests that the genetic defect in MH causes the preferential accumulation and subsequent release of norepinephrine by the sympathetic nervous system. This genetic defect may be expressed as an enzyme deficiency in converting norepinephrine to epinephrine, a defect in the active reuptake process, a defect in the enzymes that actively metabolize nor-epinephrine, or a failure in feedback inhibition.[5] This excess norepine-phrine activity may be reflected by the activation of calcium channels in skeletal muscle, with a subsequent increase in calcium ion flux in muscle cells. Allen et al. have presented data showing an increase in calcium ion activity in peroneus longus muscle of the foreleg of MHS pigs.[29]

The metabolic defect could be an actual deficiency of a functional en-zyme protein, the synthesis of an inhibitor molecule, or a deficiency of a cofactor that results in a decrease of key enzyme activity. Further studies with radioactive biogenic amines and measurement of key enzyme activi-ties in MHS pigs and human patients will be required to identify the site of the specific metabolic block.

Summary

Arterial and venous plasma samples were collected from 11 malignant hyperthermia (MHS) susceptible pigs during MH. The samples were ana-lyzed for norepinephrine and epinephrine levels by (LC-4B) electrochem-ical detection at 0.65 V, 2 nA full scale after HPLC separation on a Su-pelcosil LC-18-DB, 150 mm $\times 4.6$ mm column with a mobile phase of 0.025 M citric acid, 0.025 M Na$_2$HPO$_4$, 5×10^{-5} M EDTA, 35 mg oc-tylsodium sulfate/L, 3% MeOH, pH 3.5, at a flow rate of 1.2 ml/min. Norepinephrine plasma levels were 0.35 to 0.64 ng/ml, and epinephrine plasma levels were 0.21 to 0.59 ng/ml during the control period. Induction of MH with halothane produced a striking rise of plasma norepinephrine levels up to 40 ng/ml in arterial blood, with epinephrine levels of 24 ng/ml. One animal had 108 ng/ml of plasma norepinephrine. The plasma nor-epinephrine increases preceded the temperature rise and muscle rigor but occurred simultaneously with tachycardia. Catecholamines appear to play a key role in the pathogenesis of MH.

Acknowledgments. This work was supported in part by Organon Inc, West Orange, NJ, Grant NC45-07-83-9919. Oxmetrix, Inc, Mountain View, CA, provided the Opticath–Oxmetrix Catheter System. In addition, this in-vestigation was supported in part by the University of Missouri Institu-tional Biomedical Research Support Grant RR07053 from the National Institutes of Health and Texas Tech University Health Sciences Center Seed Research Grant.

References

1. Muhrer ME, Williams CH, Payne CG, Benedict JD (1974) Vasoconstriction and porcine hyperpyrexia. J Anim Sci 39:993
2. Williams CH, Houchins C, Shanklin MD (1975) Energy metabolism in pigs susceptible to the fulminant hyperthermia-stress syndrome. Br Med J 3:411–412 and Br Med J 1:724 (1977) errata

3. Williams CH (1976) Some observations on the etiology of the fulminant hyperthermia-stress syndrome. Perspect Biol Med 20:120–130

4. Eighmy JJ, Williams CH, Anderson RR (1978) The fulminant hyperthermia stress syndrome: Plasma thyroxine and triiodothyronine levels in susceptible and normal pigs and man, in Aldrete JA, Britt BA (eds) *Proceedings 2nd International Symposium on Malignant Hyperthermia, Denver, April, 1977.* New York: Grune & Stratton, pp 161–173

5. Nagatsu T (1973) *Biochemistry of Catecholamines: The Biochemical Method.* Baltimore: University Park Press, p 4

6. Davis TP, Gehrke CW, Gehrke CW Jr, Cunningham TD, Kuo KC, Gerhardt KO, Johnson HD, Williams CH (1978) High performance liquid chromatographic separation and fluorescence measurement of biogenic amines in plasma, urine, and tissue. Clin Chem 24:1317–1324

7. Davis TP, Gehrke CW, Gehrke CW Jr., Cunningham TD, Kuo KC, Gerhardt KO, Johnson HD, Williams CH (1979) HPLC of biogenic amines in biological materials as o–phthaldehyde derivatives. J Chromatogr 162:293–310

8. Davis TP, Gehrke CW Jr, Williams CH, Gehrke CW, Gerhardt KO (1982) Pre-column derivatization and high performance liquid chromatography of biogenic amines in blood of normal and malignant hyperthermic pigs. J Chromatogr 228:113–122

9. Williams CH, Buzello W, Dozier SE, Gehrke CW, Gerhardt KO, Wong JK (1985) Plasma levels of norepinephrine and epinephrine during malignant hyperthermia in susceptible pigs. J Chromatogr 344:71–80

10. Williams CH (1977) The development of an animal model for the fulminant hyperthermia–porcine stress syndrome. Focus on Malignant Hyperthermia, Wausau Symposium, September 13, 1974, in Henschel EO (ed) *Malignant Hyperthermia: Current Concepts.* New York: Appleton-Century-Crofts, pp 117–140

11. Buzello W, Williams CH, Chandra P, Watkins ML, Dozier SE (1985) Vecuronium and porcine malignant hyperthermia. Anesth Analg 64:515–519

12. Clark MG, Williams CH, Pfeifer WF, Bloxham DP Holland PC ,Taylor CA, Lardy HA (1973) Accelerated substrate cycling of fructose-6-phosphate in the muscle of malignant hyperthermic pigs. Nature 245:99–101

13. Williams CH, Shanklin MD, Hedrick HB, Muhrer ME, Stubbs DH, Krause GF, Payne CG, Benedict JD, Hutcheson DP, Lasley JF (1978) The fulminant hyperthermia–stress syndrome: Genetic aspects, hemodynamic and metabolic measurements in susceptible and normal pigs, in Aldrete JA, Britt BA (eds) *Proceedings 2nd International Symposium on Malignant Hyperthermia, Denver, April, 1977.* New York: Grune & Stratton, pp-113–140

14. Dozier SE, Williams CH, Ilias W (1985) Norepinephrine potentiates contracture in malignant hyperthermia susceptible (MHS) porcine muscle. Fed Proc 44(5):1376

15. Exton JH (1982) Molecular mechanisms involved in α–adrenergic responses. Trend Pharmacol Sci 3:111–115

16. Bulbring W (1946) Observations on the isolated phrenic nerve diaphragm preparation of the rat. Br J Pharmacol 1:38–61

17. Bowman WC, Nott MW (1969) Actions of sympathomimetic amines and their antagonists on skeletal muscle. Pharmacol Rev 21:27–72

18. Reuter H (1983) Calcium channel modulation by neurotransmitters, enzymes and drugs. Nature 301:569–574

19. Steiss JE, Bowen JM, Williams CH (1981) Electromyographic evaluation of malignant hyperthermia susceptible pigs. Am J Vet Res 42(7):1173–1176
20. Topel DG, Bicknell EJ, Preston KS, Christian LL, Matsushima CY (1968) Porcine stress syndrome. Mod Vet Pract 49:40–60
21. Briskey EJ (1964) Etiological and associated studies of pale, soft, exudative porcine musculature. Adv Food Res 13:89–97
22. Rosenberg H (1982) The clinical syndrome, in Brownell AKW (ed) *Transcription of the Proceedings of the Third International Workshop on Malignant Hyperthermia. September 30–October 2,* Banff, Alberta, Canada: MSI Foundation, p 4
23. Blanck TJ, Gruener RP (1983) Malignant hyperthermia. Biochem Pharmacol 32(15):2287–2289
24. Nelson TE, Flewellen EH (1983) The malignant hyperthermia syndrome. N Engl J Med 309(7):416–422
25. Campbell IT, Ellis FR, Evans RT, Mortimer MG (1983) Studies of body temperatures, blood lactate, cortisol and free fatty acid levels during exercise in human subjects susceptible to malignant hyperpyrexia. Acta Anaesth Scand 27:349–355
26. Gronert GA, Milde JH, Theye RA (1977) Role of sympathetic activity in porcine malignant hyperthermia. Anesthesiology 47:411–415
27. Lucke JN, Hall GM, Lister D (1976) Porcine malignant hyperthermia. 1. Metabolic and physiological changes. Br J Anaesth 48:297–304
28. Williams CH, Hoech GP Jr, Roberts JT (1978) Experimental malignant hyperthermia (Letter to the Editor), Anesthesiology 49:58
29. Allen P, Lopez JR, Jones D, Alamo L, Papp L, Sreter FS (1985) Measurements of $(Ca^{2+})i$ in skeletal muscle of malignant hyperthermic swine. Anesthesiology 63(3A):A268

Measurement of Biogenic Amines by HPLC-ECD: ECD as a Diagnostic Tool for Malignant Hyperthermia

Klaus O. Gerhardt, Charles W. Gehrke,
Charles H. Williams, and Joseph K. Wong

Introduction

Maximum precision and accuracy of measurements are essential in obtaining information on data for monitoring biologically important molecules, for example, catecholamines, in biologic fluids as related to biologic changes with the lethal malignant hyperthermia (MH) syndrome. Reversed phase high performance liquid chromatography (RP–HPLC) is the most widely used tool in the separation of these sensitive and labile molecules, coupled most commonly with either the very selective and sensitive fluorometric measurement[1-4] or electrochemical detection (ECD),[5-7] pioneered by Kissinger, for quantitation. The need for the simultaneous measurement of norepinephrine and epinephrine in very small samples made the sensitive ECD the method of choice. However, many endogenous substances in plasma respond electrochemically as interfering peaks, making a sample cleanup necessary. The instability of the catecholamines and their high water solubility do not favor an efficient solvent extraction. A cleanup method of catecholamines developed by Anton and Sayre[8] has been modified.[7] The procedure is based on the adsorption of catecholamines on alumina at an alkaline pH, the removal of water–soluble matrix components and the desorption of catecholamines from the alumina at acidic pH with $HClO_4$. We have recently modified and optimized the HPLC conditions coupled with ECD to measure quantitatively the norepinephrine and epinephrine levels in plasma during the development of the MH syndrome. A detailed study of the hemodynamic parameters, body temperature changes, and other physiologic factors has been published separately.[10,11]

Malignant hyperthermia is a hereditary disorder, and the syndrome also can be induced through the administration of some anesthetics and depolarizing muscle relaxants in susceptible animals, although most cases of MH triggered by an anesthetic have been reported with halothane.[12]

It is found not only in pigs and other animals but also in humans.[13] The

MH susceptible (MHS) pig provides a valuable animal model for obtaining information that can be used in studying and understanding the MH syndrome in humans, which could lead to the prevention and treatment of the genetic disease.

Methods and Materials

Chemicals

All reagents used were of the highest available purity. The calibration standards norepinephrine, epinephrine, and the internal standard (IS) 3,4–dihydroxybenzylamine (DHBA) were obtained from Sigma Chemical Co., St. Louis, MO. The 0.1 M perchloric acid solutions of these standards were stored at 4°C for daily use.

For the mobile phase, we used citric acid from J.T. Baker Chemical Co., Phillipsburg, NJ, Na_2HPO_4 and EDTA from Sigma Chemical Co., St. Louis, MO, sodium octyl sulfate from Eastman Kodak Chemical Co., Rochester, NY, and methanol, glass distilled, from MCB, Division of EM Science Industries, Inc., Gibbstown, NJ, perchloric acid, from Fisher Scientific Co., St. Louis, MO, and nanopure double distilled water.

Animals

Malignant hyperthermia susceptible pigs were raised from our genetic strain of halothane–sensitive pigs.[12] Each animal ($n = 11$) was tested with halothane at 8 to 10 weeks of age to determine MH susceptibility.[12] The animals were fed a 16% protein swine ration and used for these experiments at 52 to 62 kg body weight.

Chromatography

The levels of norepinephrine and epinephrine in the acidic phase after alumina cleanup were measured following separation by RP-HPLC coupled with highly sensitive ECD. A model U6-K valve injector (Waters Associates, Milford, MA), a Technicon Fast-LC pump (Technicon Instruments, Tarrytown, NY), and an LC-4B electrochemical detector (Bioanalytical Systems, West Lafayette, IN) were the instrument components used. The analytical column used was Supelcosil LC-18-DB (Supelco, Inc., Bellefonte, PA), 5 μm, base deactivated, 150 mm × 4.6 mm, protected by a Supersorb C_{18}-guard column. The mobile phase consisted of 0.025 M Na_2HPO_4, 0.025 M citric acid, 5×10^{-5} M EDTA, 35 mg sodium octyl sulfate/L and 3% MeOH, pH 3.5. The flow rate was maintained at 1.2 ml/minute, and the column temperature at 31°C. The EC detector was equipped with a glassy carbon electrode and set at +0.65 V versus the Ag/AgCl reference electrode, and the response was 2 nA full scale.

Sample Preparation

Five-milliliter blood samples were drawn from the femoral artery, external iliac vein, and pulmonary artery at three stages: the control period, at the onset of tachycardia after administration of 2% halothane, and then at 5 to 10 minute intervals. Immediately after the blood was taken, each blood sample was mixed with a freshly prepared EDTA-sodium metabisulfite solution (20 g EDTA and 10 g sodium metabisulfite/L) while kept in an ice water bath, and $HClO_4$ was added for deproteination of the plasma as previously reported.[1-3,14] The samples were stored immediately at $-70°C$ in darkness.

The alumina cleanup method[7] allows the removal of numerous compounds interfering with the HPLC ECD and the measurement of catecholamines. The plasma samples were thawed at room temperature, and an exactly measured 0.5 ml aliquot was pipetted into a 12-ml conical centrifuge tube containing 50 mg alumina, then exactly measured 20 ng DHBA internal standard (IS) was added. The suspension was gently mixed, and the pH was adjusted to 8.6 with 1 M TRIS buffer containing 2×10^{-5} M EDTA, followed by vigorous shaking of the suspension for 10 minutes to adsorb the catecholamines on the alumina. After the alumina settled, the wall of the tube was rinsed with 3 ml H_2O to collect all alumina at the

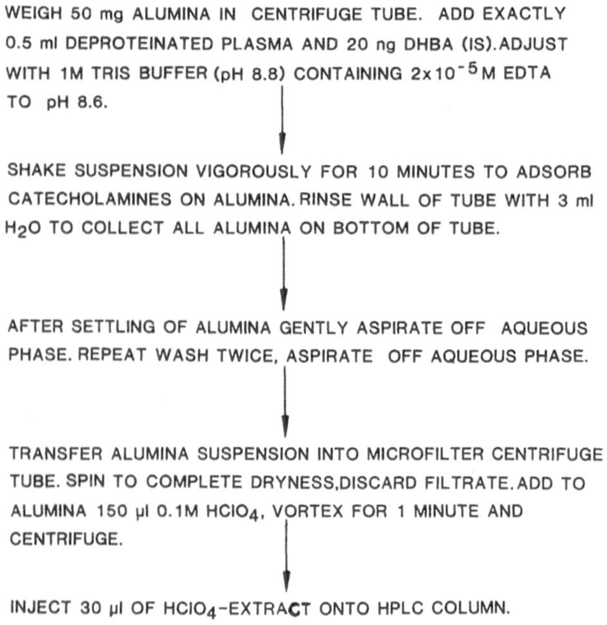

WEIGH 50 mg ALUMINA IN CENTRIFUGE TUBE. ADD EXACTLY
0.5 ml DEPROTEINATED PLASMA AND 20 ng DHBA (IS). ADJUST
WITH 1M TRIS BUFFER (pH 8.8) CONTAINING 2×10^{-5} M EDTA
TO pH 8.6.

SHAKE SUSPENSION VIGOROUSLY FOR 10 MINUTES TO ADSORB
CATECHOLAMINES ON ALUMINA. RINSE WALL OF TUBE WITH 3 ml
H_2O TO COLLECT ALL ALUMINA ON BOTTOM OF TUBE.

AFTER SETTLING OF ALUMINA GENTLY ASPIRATE OFF AQUEOUS
PHASE. REPEAT WASH TWICE, ASPIRATE OFF AQUEOUS PHASE.

TRANSFER ALUMINA SUSPENSION INTO MICROFILTER CENTRIFUGE
TUBE. SPIN TO COMPLETE DRYNESS, DISCARD FILTRATE. ADD TO
ALUMINA 150 µl 0.1M $HClO_4$, VORTEX FOR 1 MINUTE AND
CENTRIFUGE.

INJECT 30 µl OF $HClO_4$-EXTRACT ONTO HPLC COLUMN.

FIGURE 3.1. Plasma sample cleanup method of catecholamines by alumina adsorption prior to HPLC analysis

bottom of the tube, aqueous phase was gently aspirated off, and the alumina was washed twice with 3 ml H_2O and the supernatant discarded. About 1 ml water was added, the slurry was transferred into a microfilter centrifuge tube (Bioanalytical Systems) and centrifuged at 3000 g for 30 seconds, and the aqueous filtrate was discarded. To the alumina were added 150 μL of 0.1 M $HClO_4$ to desorb the catecholamines. After mixing (vortex) of the slurry for 1 minute and spinning at 3000 g for 1 minute, an aliquot of 30 μL of the $HClO_4$ extract was injected onto the HPLC.

Results

Most important in the analysis of catecholamines is to protect these molecules from chemical breakdown, beginning with the blood sampling and continuing throughout the analytical procedure. We have found that EDTA and sodium metabisulfite are effective in preserving the catecholamines.[1,2] Prior to analysis, the samples were stored at $-70°C$ in the dark.

The extraction method of norepinephrine and epinephrine using alumina adsorption and desorption, as outlined in Figure 3.1 was very reproducible and accurate, as shown in Table 3.1. A very good precision of $\pm$ 1.66% relative standard deviation (RSD) for norepinephrine and $\pm$ 2.06% for epinephrine (RSD) was achieved based on the average of 15 independent analyses performed on different days. The precision values document the high reproducibility of our catecholamine measurements. The extraction efficiency of the alumina adsorption method was very high for both norepinephrine and epinephrine, as expressed in the very good recovery values of 99.2% and 96.0% respectively. The recovery values are the average of 15 additions of norepinephrine and epinephrine to pooled plasma samples that were analyzed independently on different days.

Figure 3.2 describes the HPLC separation of a standard mixture of norepinephrine, epinephrine and DHBA (IS) with ECD at the nanogram level. The detection limit is at 20 to 50 pg of norepinephrine and epinephrine per ml plasma at an applied potential of $+0.65$ V against an Ag/AgCl reference electrode. The separation was applied to the study of pigs af-

TABLE 3.1. Recovery, precision, and detection limits of norepinephrine and epinephrine from pig plasma

Compound	Recovery (%)[a]	Precision (%)[b]	Detection Limit (pg/ml plasma)
Norepinephine	99.2	$\pm$1.66	20–50
Epinephrine	96.0	$\pm$2.06	20–50

[a]Average of 15 additions to pooled pig plasma, analyzed independently on different days.
[b]Average of 15 independent analyses on different days.

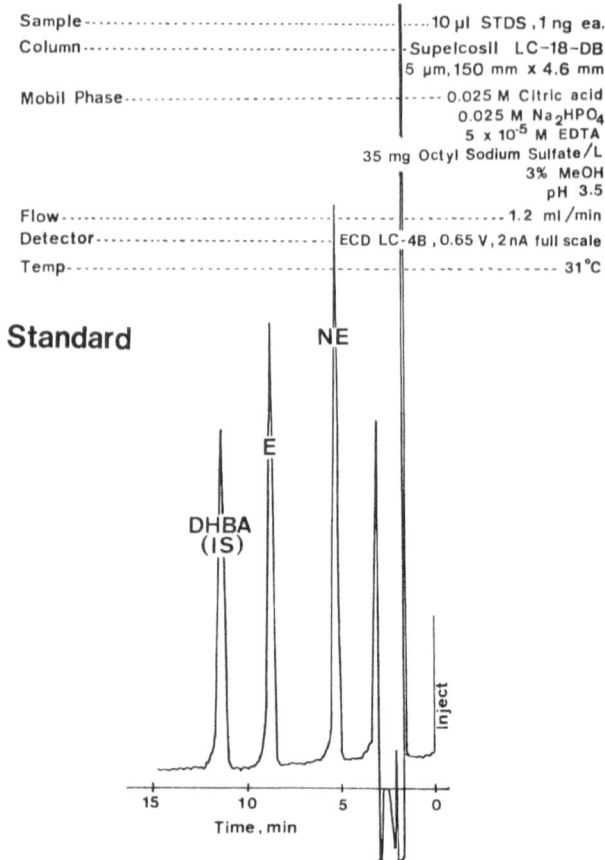

Sample .... 10 μl STDS ,1 ng ea.
Column .... Supelcosil LC-18-DB
                                                  5 μm, 150 mm x 4.6 mm
Mobil Phase .... 0.025 M Citric acid
                                                  0.025 M Na₂HPO₄
                                                  5 x 10⁻⁵ M EDTA
                                                  35 mg Octyl Sodium Sulfate/L
                                                  3% MeOH
                                                  pH 3.5
Flow .... 1.2 ml /min
Detector .... ECD LC-4B , 0.65 V, 2 nA full scale
Temp .... 31°C

Standard   NE

                       E

              DHBA
              (IS)

                              Inject

15   10   5   0
         Time , min

FIGURE 3.2. Reversed phase HPLC separation of a standard mixture of norepinephrine and epinephrine and measurement by coupled ECD. Sample, 10 uL standards, 1 ng each; column, Supelcosil LC–18–DB, 5 um, (150 mm x 4.6 mm); mobile
phase, 0.025 M Na2HPO4, 0.025 M citric acid, 5 x 10-5 M EDTA, 35-g sodium
octyl sulfate/L, 3% (v/v) MeOH, and pH 3.5; flow rate, 1.2 ml/min; temperature,
31oC; detector, ECD (LC-4B), +0.65 V vs Ag/AgCl reference, 2nA full scale;
internal standard (IS), 3,4–dihydroxybenzylamine (DHBA). Peaks: NE, norepinephrine; E, epinephrine.

flicted with the genetic disorder of MH. Figure 3.3 presents the RP HPLC
separation of the catecholamines norepinephrine, epinephrine, and DHBA
(IS) of an MHS control pig. The plasma levels of norepinephrine during
the control period ranged from 0.35 ng/ml to 0.64 ng/ml and 0.20 ng/ml
to 0.6 ng/ml for epinephrine. The administration of halothane induced the
MH syndrome, leading to tachycardia and in some pigs to muscle rigor.

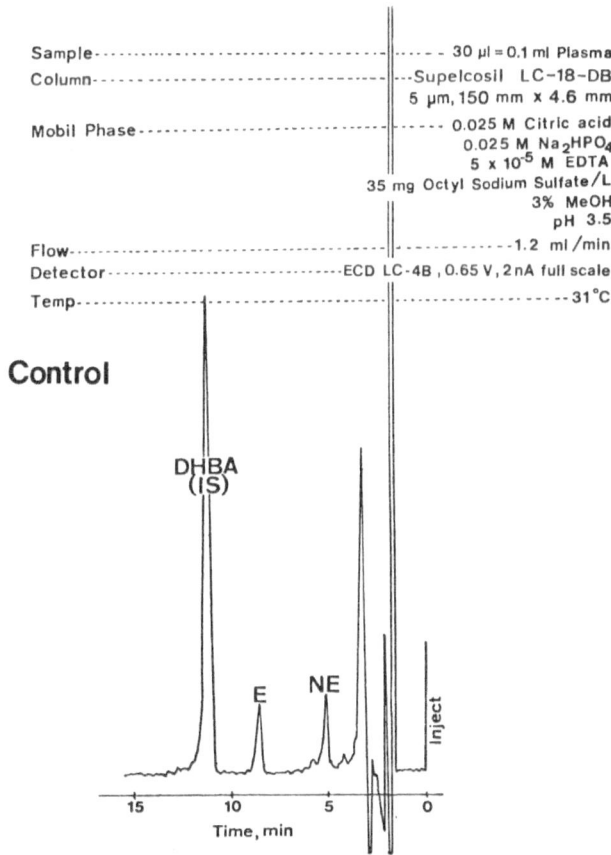

Sample - | - - - - - 30 µl = 0.1 ml Plasma
Column - | - - - Supelcosil LC-18-DB
                                                                       5 µm, 150 mm x 4.6 mm
Mobil Phase - | - - - - - 0.025 M Citric acid
                                                                         0.025 M Na₂HPO₄
                                                                          5 x 10⁻⁵ M EDTA
                                                      35 mg Octyl Sodium Sulfate/L
                                                                               3% MeOH
                                                                                pH 3.5
Flow - | - - - - - - - - - - - -1.2 ml /min
Detector - ECD LC-4B , 0.65 V, 2 nA full scale
Temp - - - - - - - - - - - - - - - | - - - - - - - - - - - - - - - | - - - - - - - - - - - - - -31°C

Control

DHBA
(IS)

E   NE

Inject

15   10   5   0
Time, min

FIGURE 3.3. Reversed phase HPLC separation of norepinephrine and epinephrine and measurement by ECD in plasma from a control pig. Sample: 30 µL equivalent to 0.1 ml plasma. All other conditions as in Figure 3.2. (From J. Chromatogr. 344:75–80, 1985. Reprinted with permission of Elsevier.)

Soon after the anesthetic was given, a massive release of up to 108 ng norepinephrine/ml of plasma was measured, and the values for epinephrine increased to about 40 ng/ml. Figure 3.4 presents the significantly increased norepinephrine and epinephrine levels during the MH syndrome. In Figure 3.5 the changes in norepinephrine and epinephrine levels of a pig during halothane-induced MH are shown. This animal also developed muscle rigor. The highest levels of norepinephrine and epinephrine in plasma from this pig at the same time were 85 ng and 40 ng/ml, respectively. Unrelated pigs of a different breed exposed to halothane did not develop the syndrome, and no significant changes in norepinephrine and epinephrine were measured.[10,14]

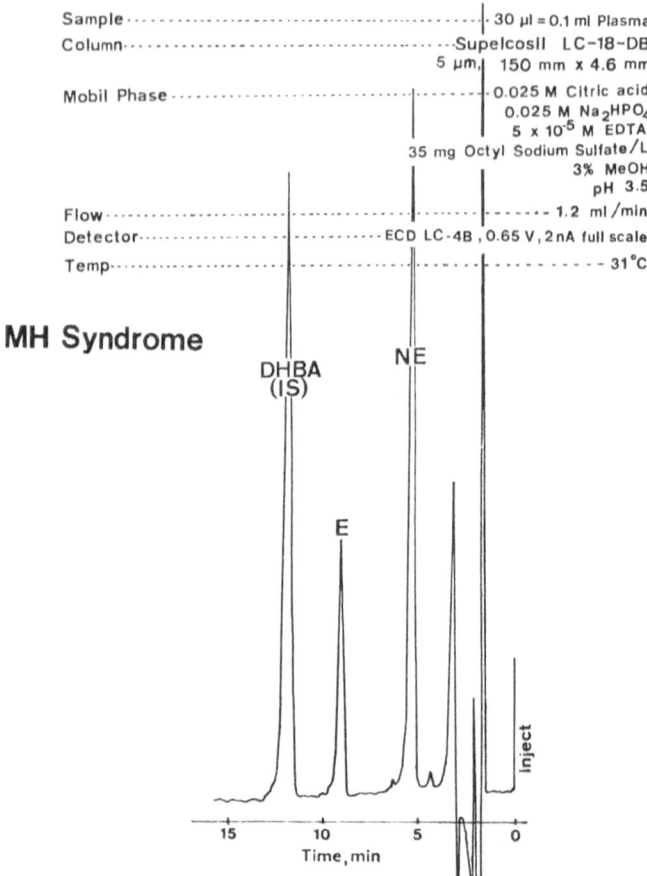

Sample · 30 μl = 0.1 ml Plasma
Column · Supelcosll LC–18–DB
                                                5 μm, 150 mm x 4.6 mm
Mobil Phase · 0.025 M Citric acid
                                          0.025 M Na₂HPO₄
                                              5 x 10⁻⁵ M EDTA
                              35 mg Octyl Sodium Sulfate/L
                                                      3% MeOH
                                                        pH 3.5
Flow · 1.2 ml/min
Detector · ECD LC-4B , 0.65 V, 2 nA full scale
Temp · – 31°C

MH Syndrome

DHBA
(IS)

NE

E

Inject

15   10   5   0
Time, min

FIGURE 3.4. Reversed phase HPLC separation of norepinephrine and epinephrine and measurement by coupled ECD in plasma from a pig during MH syndrome. All other conditions as in Figure 3.2.

Discussion

The application of the presented HPLC-ECD method enabled the precise and accurate analysis of norepineprhine and epinephrine simultaneously in pig plasma. Our data provided evidence that during the start of tachycardia, which heralds the development of the MH syndrome, significant increases in norepinephrine and epinephrine levels occur. Thus, the norepinephrine release and the increased heart rate are two simultaneously occurring events.[10] The overall severity of the syndrome may be a reflection of the initial amount of norepinephrine released and then accumulating as the syndrome progresses.

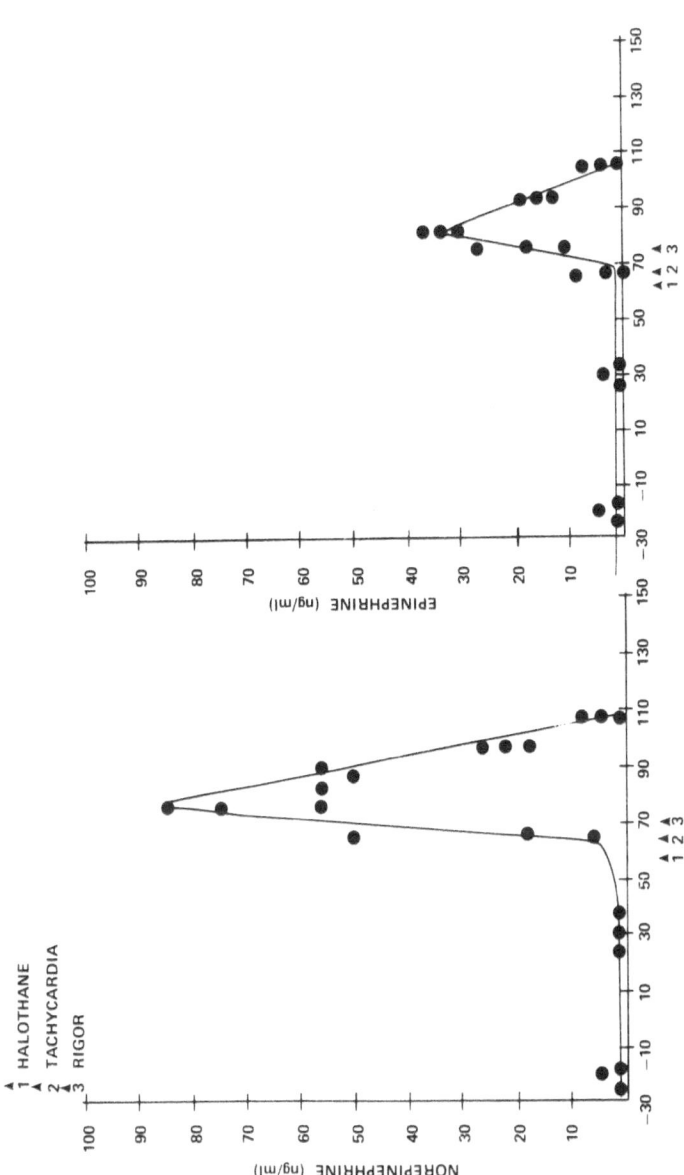

FIGURE 3.5. Plasma levels of norepinephrine and epinephrine prior to and during halothane-induced MH. The pig developed muscle rigor also. Each point represents independently measured catecholamine levels from serial samples of arterial, pulmonary artery, and venous plasma samples. (From J. Chromatogr. 344:75–80, 1985. Reprinted with permission of Elsevier.)

Analogously, tachycardia has been recognized as a symptom of MH in human patients.[10] Based on our experimental animal data, a study of biogenic amine levels and patterns in human MH patients would be important in identifying susceptible people, thus leading to treatment and possible prevention of the syndrome.

Summary

Plasma levels of norepinephrine and epinephrine were measured during HPLC with electrochemical detection (ECD) to monitor the release of these biogenic amines before and during the halothane induced MH syndrome in susceptible pigs. The presented method is simple and sensitive, allowing the measurement of norepinephrine and epinephrine with very good precision, ranging from 1.7 to 2.1% RSD, and high accuracy, reflected in recoveries of 99.2% for norepinephrine and 96% for epinephrine. The detection limit of norepinephrine and epinephrine is 20 to 50 pg/ml plasma. Arterial and venous plasma samples were analyzed for norepinephrine and epinephrine by HPLC-ECD at +0.65V (glassy carbon electrode) against an Ag/AgCl reference electrode. Good separation was achieved on a C18-column of 5 μm particle size with a mobile phase of 0.025 M Na_2HPO_4, 0.025 M citric acid, 5×10^{-5} M EDTA, 3% MeOH, and 35 mg sodium octyl sulfate/L, pH 3.5 at 31°C. During the control period, plasma levels of norepinephrine and epinephrine were about 0.5 ng/ml and 0.2 ng/ml, respectively. As MH developed values of up to 108 ng/ml for norepinephrine and 40 ng/ml for epinephrine were measured. The data provide further evidence for their involvement in MH.

Acknowledgments. This research was supported in part by a Biomedical Research Support Grant RR07053, by PHS Grant 5R01CA28668-03, Grant NC 45-07-83-9919 from Organon Inc., West Orange, NJ, and by Texas Tech University Health Sciences Center Institute for Biomedical Research Seed Grant. We thank Jennifer M. Welch for technical assistance.

References

1. Davis TP, Gehrke CW Jr, Cunningham TD, Kuo KC, Gerhardt KO, Johnson HD, Williams CH (1978) High-performance liquid chromatographic separation and fluorescence measurement of biogenic amines in plasma, urine and tissue. Clin Chem 24:1317–1324
2. Davis TP, Gehrke CW, Gehrke CW Jr, Cunningham TD, Kuo KC, Gerhardt KO, Johnson HD, Williams CH (1979) High-performance liquid chromatographic analysis of biogenic amines in biological materials as *o*–phthalaldehyde derivatives. J Chromatogr 162:293–310
3. Davis TP, Gehrke CW Jr, Gerhardt KO, Williams CH, Gehrke CW (1984) Pre-column derivatization, HPLC, and fluorescence measurement of biogenic

amines in biological materials, in Kabra PM, Marton LJ (eds) *Clinical Liquid Chromatography, Analysis of Endogenous Compounds*. Boca Raton, FL: CRC Press, vol. 2, pp 53–70

4. Krstulovic AM (1982) Review–Investigations of catecholamine metabolism using HPLC. Analytical methodology and clinical applications. J Chromatogr 229:1–34

5. Kissinger PT, Riggin RM, Alcorn RL, Rau LD (1975) Estimation of catecholamines in urine by high performance liquid chromatography with electrochemical detection. Biochem Med 13:299–306

6. Davis GC, Shoup RE, Kissinger PT (1981) Strategies for determination of serum or plasma norepinephrine by reverse-phase liquid chromatography. Anal Chem 53:156–159

7. Shoup RE, Meyer GS (1984) Determination of plasma norepinephrine by liquid chromatography/electrochemistry, in Kabra PM, Marton LJ (eds) *Clinical Liquid Chromatography, Analysis of Endogenous Compounds*. Boca Raton, FL: CRC Press, pp 47–51

8. Anton AH, Sayre DF (1962) A study of the factors affecting the aluminum oxide-trihydroxyindole procedure for the analyses of catecholamines. J Pharmacol Exp Ther 138:360–375

9. Lee PJ (1983) Masters thesis: Biologic Markers for Melanoma. Columbia, MO: University of Missouri

10. Williams CH, Dozier SE, Buzello W, Gehrke CW, Wong JK, Gerhardt KO (1985) Plasma levels of norepinephrine and epinephrine during malignant hyperthermia in susceptible pigs. J Chromatogr 344:71–80

11. Buzello W, Williams CH, Chandra P, Watkins ML, Dozier SE (1985) Vecuronium and porcine malignant hyperthermia. Anesth Analg 64:515–519

12. Williams CH (1977) Focus on malignant hyperthermia–Wasau Symposium, September 13, 1974, in Henschel EO (ed) *Malignant Hyperthermia: Current Concepts*. New York: Appleton–Century–Crofts, pp 117–140

13. Rosenberg H (1982) The clinical syndrome, in *Transcription of the Proceedings of the Third International Workshop on Malignant Hyperthermia, September 30–October 2*. Banff, Alberta, Canada, p 4

14. Davis TP, Gehrke CW Jr, Williams CH, Gehrke CW, Gerhardt KO (1982) Pre-column derivatization and high performance liquid chromatography of biogenic amines in blood of normal and malignant hyperthermic pigs. J Chromatogr 228:113–122

Hemodynamics in Malignant Hyperthermia Susceptible Pigs During Malignant Hyperthermia

Charles H. Williams, Susan E. Dozier,
and Mary Farias

Introduction

Several years ago, we observed the stress induced malignant hyperthermia (MH) syndrome in our malignant hyperthermia susceptible (MHS) pigs after transport. We were astounded by the cardiac palpitations that were readily apparent in the arterial supply to the ears and the carotid or jugular throbbing visible in the neck region whenever spontaneous MH developed. In white MHS pigs, we could observe the cyanotic appearance of the ears, nose, and skin. It became apparent from our calorimeter studies that an intense peripheral vasoconstriction was the lethal part of the acute MH syndrome.[1-4] Our previous studies further documented the acute hemodynamic shutdown that produced cardiovascular collapse in inbred MH pigs.[1,2] We have pursued the hemodynamic studies in order to document further the hemodynamic response to developing MH in a larger population of MHS pigs of both inbred and outcross breeding patterns.

Methods and Materials

Malignant hyperthermia susceptible pigs were raised from our MHS breeding colony maintained at the Crawford Rd. Experimental Medicine Research Farm. The pigs were challenged by mask with 4% halothane for 3 minutes, then maintained on 2% halothane for up to 20 minutes to determine their susceptibility. Positive reactors to halothane were reserved for these experiments. Twelve MHS pigs were transported to the laboratory and allowed access to water but were fasted overnight in preparation for the surgical procedures. Body weights ranged from 20 to 30 kg. Thiopental (22 mg/kg) was given intraperitoneally. An ear vein was cannulated, and 0.9% NaCl was used as the carrier for intermittent doses of thiopental as needed. The animal was intubated and mechanically ventilated at a rate to give an end tidal CO_2 reading of 42 torr at STP.

Invasive monitoring consisted of an Oximetrix Opticath® catheter inserted at the femoral vein and advanced to the wedge position in the pulmonary artery. The femoral artery was cannulated with a 16G catheter for arterial pressure recording.

Other monitoring consisted of heart rate, ECG, core and rectal temperature, cardiac output by thermal dilution with a model 9520 Edwards cardiac output computer, central venous pressure (CVP), pulmonary artery pressure (PAP), and pulmonary capillary wedge pressure (PCWP). Malignant hyperthermia was induced by administering 2% halothane in the rebreathing circuit.

All data calculations were performed on our Hewlett-Packard model 86 computer. Statistical analyses (t-test or Wilcoxson ranked sum test) were performed to assess the statistical significance of the data.

Results

Table 4.1 shows the blood pressure data collected from 35 normal pigs and 35 MHS pigs in our hemodynamics laboratory during the control period after all invasive lines were installed and before induction of the MH syndrome. Blood pressure was not significantly different ($P < 0.44$) in MHS pigs (mean 110.62 mm Hg) than in normal pigs (mean 106.57 mm Hg).

The raw data in Figure 4.1 illustrate the kind of MHS pig that has a very slow development of MH, as evidenced by the slow decrease of venous O_2 saturation and the long exposure period (over 1 hour) to halothane. Note that there were increased dysrhythmias of the heart as reflected by the wide variation in recorded heart rate.

The raw data in Figure 4.2 illustrate the rapid development of MH in a highly susceptible pig after only 4 minutes of halothane exposure. Note that this pig was hypertensive during the control period (320/280 torr), there was a drop in mean arterial pressure when halothane was started, and then the mean arterial pressure exceeded 400 torr when the halothane was turned off. The extremely high blood pressure was maintained until cardiovascular collapse occurred.

Figure 4.3 shows that oxygen consumption increased fourfold as the MH syndrome progressed. Cardiac index (Figure 4.4) was decreased slightly after halothane administration but was not significantly affected by the MH syndrome. Stroke volume index (Figure 4.5) decreased significantly at halothane administration and remained decreased during MH.

Heart rate (Figure 4.6) was significantly elevated 5 minutes before the appearance of any other physiologic change, including change in body temperature. Total peripheral resistance (Figure 4.7) decreased as the MH syndrome developed in these outcross pigs.

Left ventricular stroke work index (Figure 4.8) was significantly decreased on halothane administration, remained significantly lower during the tachycardia phase, increased as MH developed, and then decreased again as the cardiovascular collapse ensued.

TABLE 4.1. Blood pressure data collected from 35 normal pigs and 35 MHS pigs during control period after all invasive lines were installed

	Normal Sample 1 $n = 35$	MHS Sample 2 $n = 35$
1	83.0	79.0
2	108.0	97.0
3	95.0	101.0
4	105.0	117.0
5	84.0	65.0
6	120.0	107.0
7	100.0	133.0
8	90.0	90.0
9	109.0	87.0
10	114.0	111.0
11	97.0	81.0
12	98.0	96.0
13	80.0	91.0
14	75.0	58.0
15	90.0	88.0
16	84.0	110.0
17	93.0	121.0
18	93.0	130.0
19	99.0	121.0
20	87.0	195.0
21	133.0	92.0
22	130.0	90.0
23	137.0	126.0
24	97.0	106.0
25	123.0	135.0
26	120.0	107.0
27	127.0	119.0
28	117.0	105.0
29	120.0	117.0
30	123.0	117.0
31	113.0	147.0
32	113.0	133.0
33	133.0	160.0
34	123.0	123.0
35	117.0	117.0
Mean	106.57 ± 17.21	110.62 ± 26.37

There was not a significant difference ($P < 0.44$) in the blood pressure of MHS (mean 110.62 mm Hg) versus normal pigs (mean 106.57 mm Hg), evaluated using unpaired t-test.

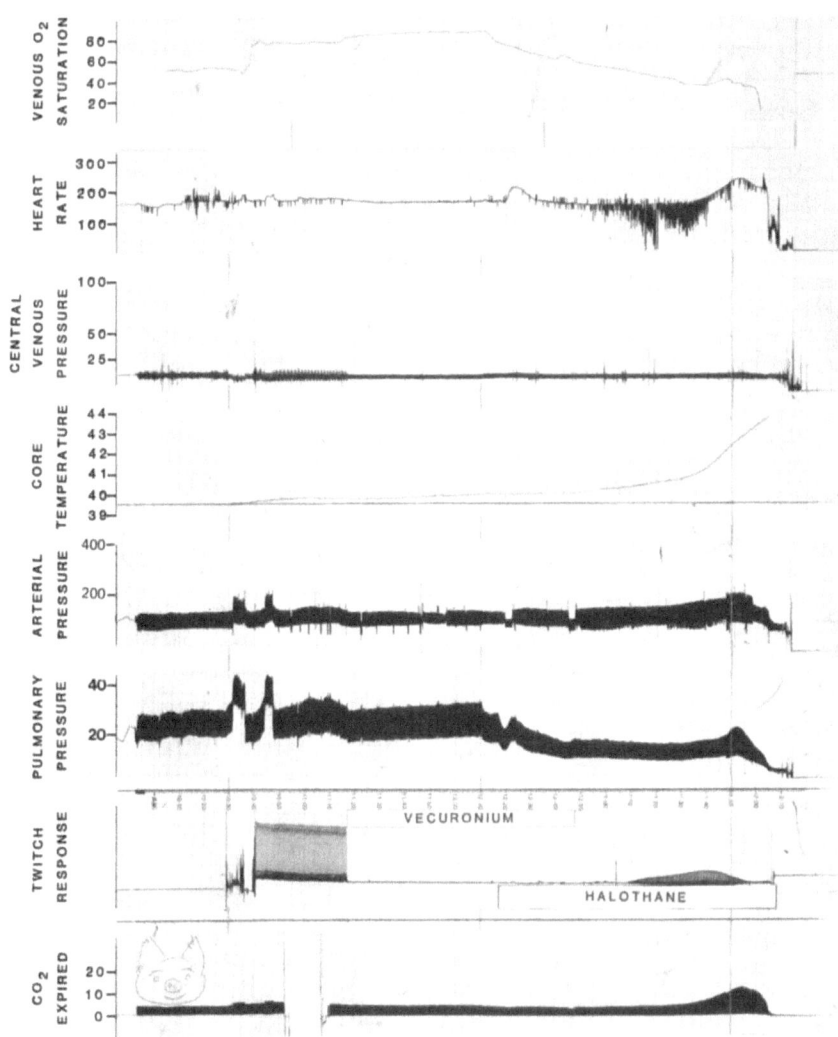

FIGURE 4.1. Typical in vivo recordings in an MHS pig with very slow development of MH. Halothane exposure was continued for over 1 hour.

Pulmonary vascular resistance (Figure 4.9) decreased slightly on halothane administration, then increased markedly as tachycardia developed. These changes were not statistically significant, but two pigs developed pulmonary edema.

Figure 4.10 illustrates the development of a core temperature of 45.20°C in an MHS pig with muscle rigor. Figure 4.11 illustrates the development of a core temperature of 44.31°C in an MHS pig without muscle rigor.

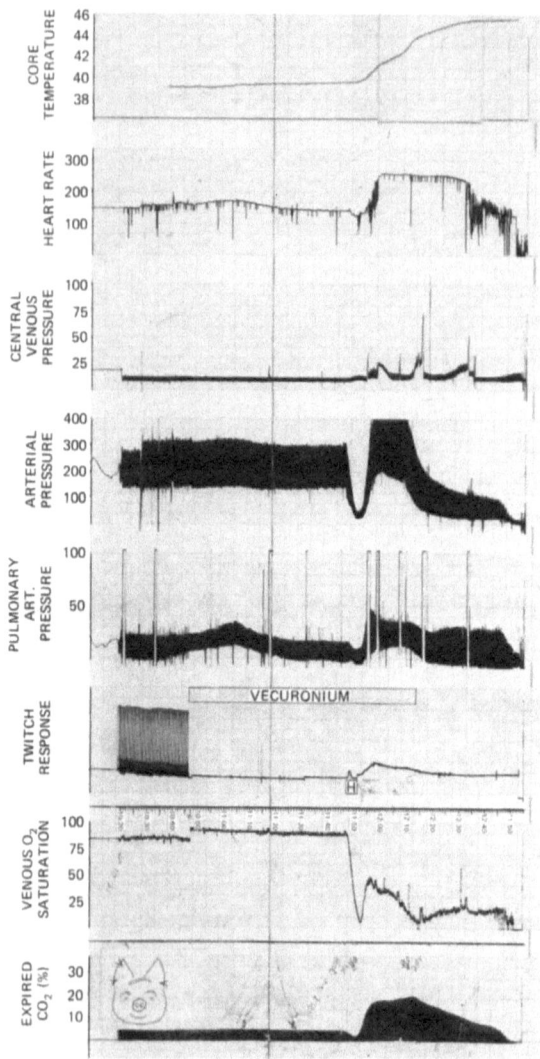

FIGURE 4.2. Typical in vivo recordings in an MHS pig with very rapid development of MH. Total halothane exposure was for 4 minutes.

Discussion

The development of a seriously compromised cardiovascular system with tachycardia, cardiac dysrhythmias, an intense peripheral vasoconstriction, and a rapid cardiovascular collapse are the hallmarks of the development of lethal MH in susceptible pigs. Similar results and conclusions

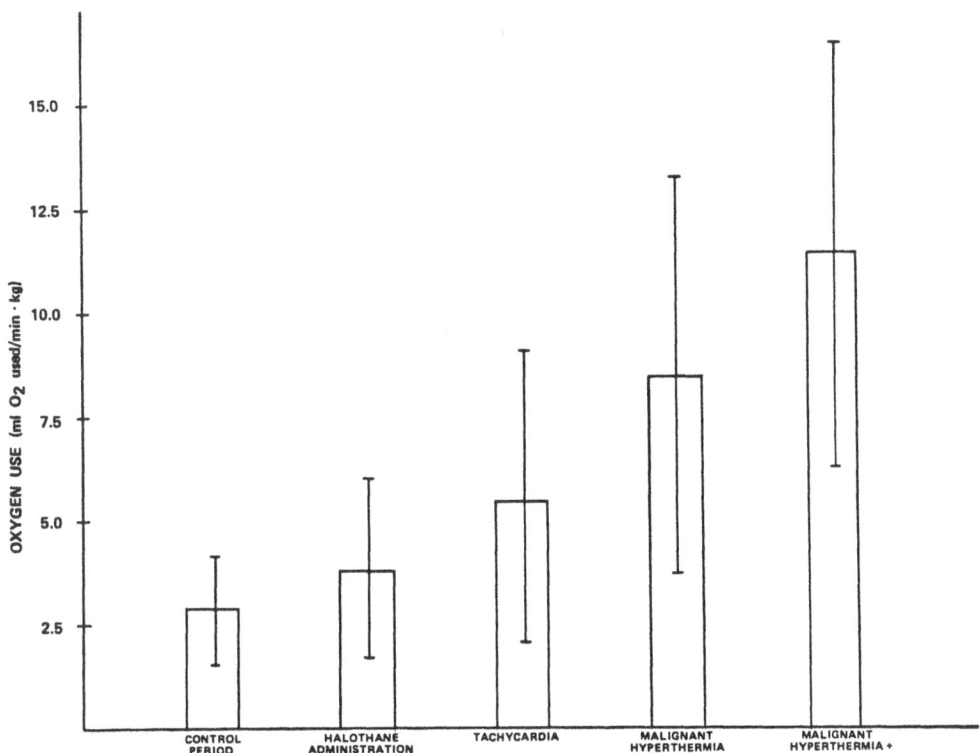

FIGURE 4.3. Oxygen utilization during the development of MH as measured by arterial/venous differences.

can be drawn from human MH case reports.[5] In one experiment (unpublished results), we had an immediate cardiac arrest after injecting succinylcholine without the development of muscle rigor or a temperature rise in an MHS pig. Previous studies of hemodynamics during MH in susceptible swine have presented limited data on blood pressure, cardiac output, and other hemodynamic variables.[6-15]

The outcross MHS pigs we used in this series of experiments had a decreased peripheral resistance as the MH syndrome developed. This response contrasts markedly with the increased peripheral resistance we observed previously in our inbred MHS pigs[1,2] and illustrates very graphically the role that the dosage of the genetic defect plays in determining the pathophysiologic responses in affected animals. Several human patients have exhibited increased peripheral resistance especially in the hands and feet.[16,17]

Tachycardia is the first external physiologic response of impending MH. Tachycardia occurs 5 minutes before any other symptom of MH. These data confirm our earlier observations on the development of tachycardia

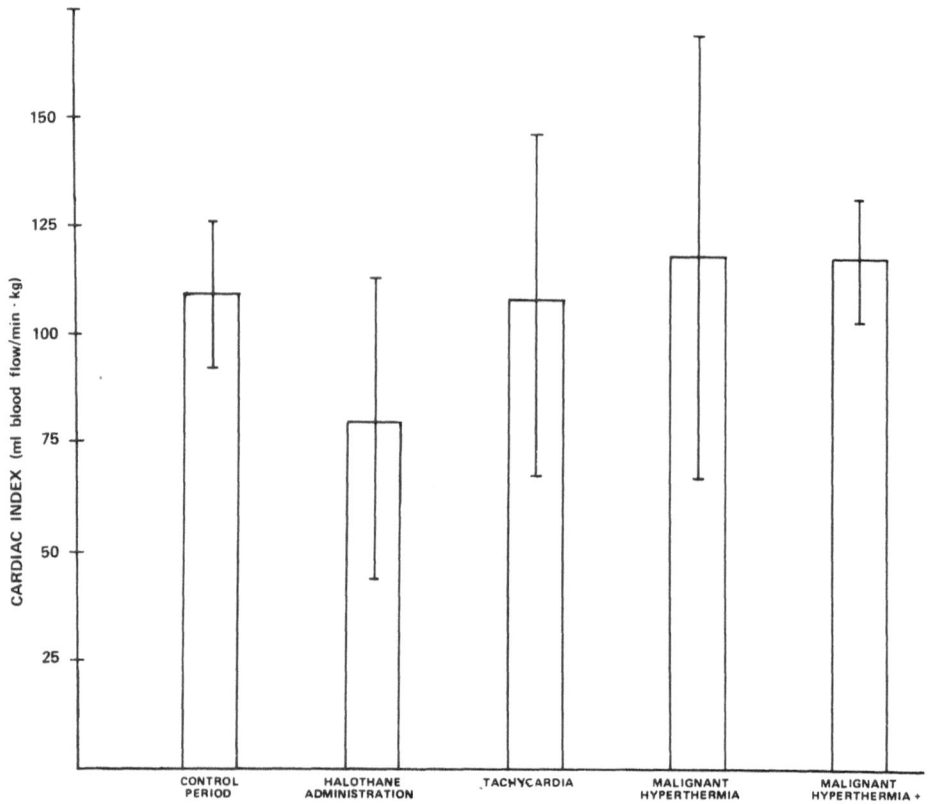

FIGURE 4.4. Cardiac index during the development of MH.

and MH in pigs treated with pharmacologic doses of metubine iodide.[18] The development of tachycardia may be used clinically as an indication of impending MH in human patients.

Increased pulmonary vascular resistance can lead to pulmonary edema. The increased pulmonary vascular resistance (although not statistically significant) we have measured may be an important component in the development of spontaneous MH in susceptible pigs. The spontaneous development of MH in conscious pigs results in a labored, open-mouthed breathing, reflecting a failure to obtain sufficient oxygen via the pulmonary vascular system. We have observed that 95% to 98% arterial O_2 saturation can be maintained during MH by providing a 100% oxygen supply. However, venous O_2 saturation approaches zero during MH. The marked A/V difference is a strong indication of MH and could be used clinically to diagnose the syndrome.

The development of MH with muscle rigor or without muscle rigor needs to be rationalized and discussed. Twenty-five percent of human MH pa-

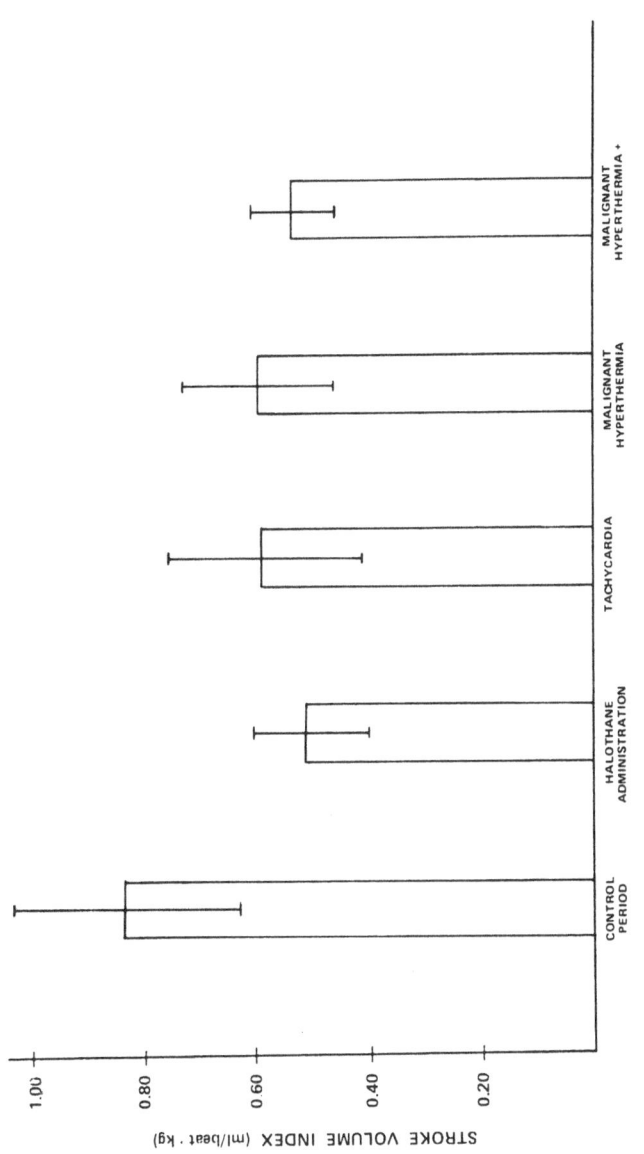

FIGURE 4.5. Stroke volume index during the development of MH.

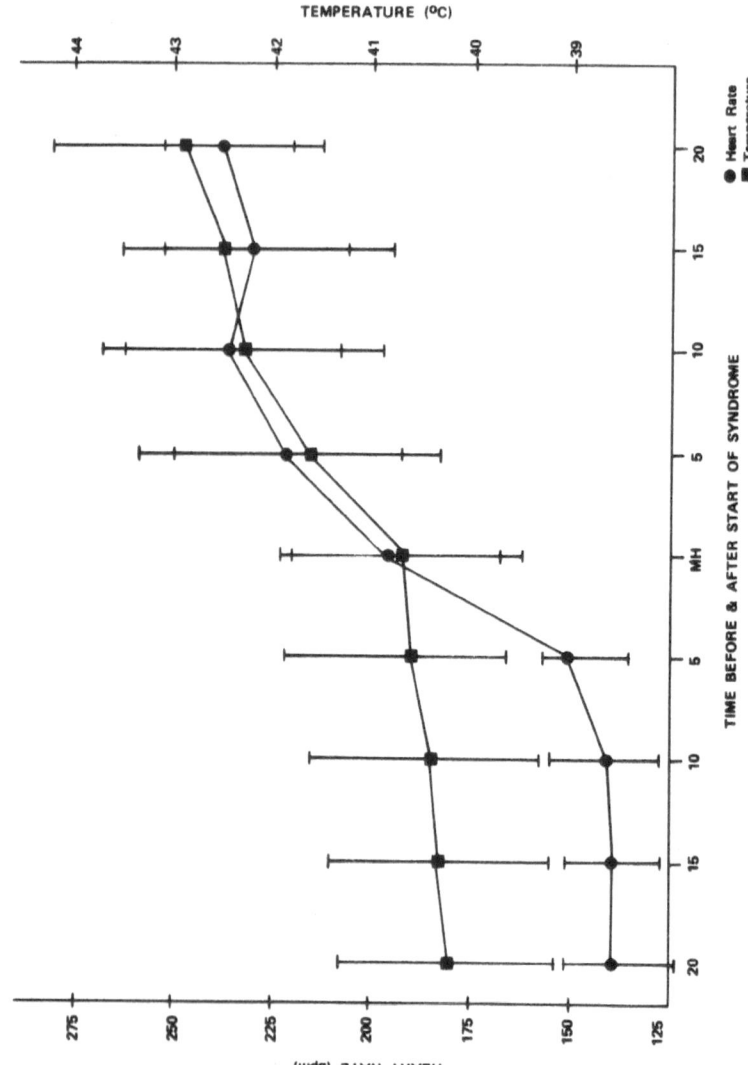

FIGURE 4.6. Core temperature and heart rate during the development of MH.

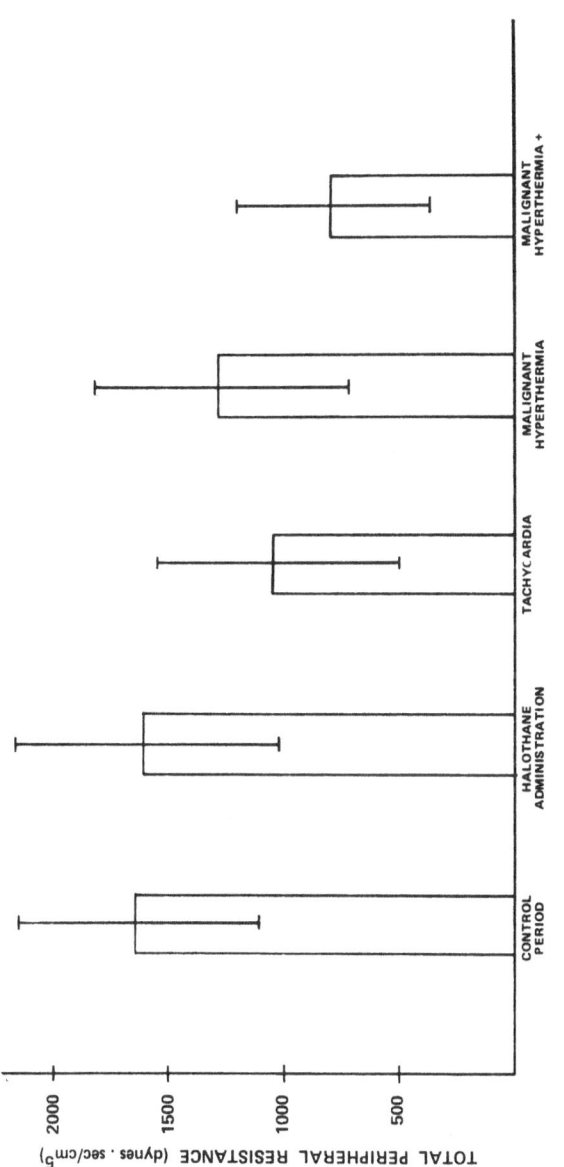

FIGURE 4.7. Total peripheral resistance during the development of MH.

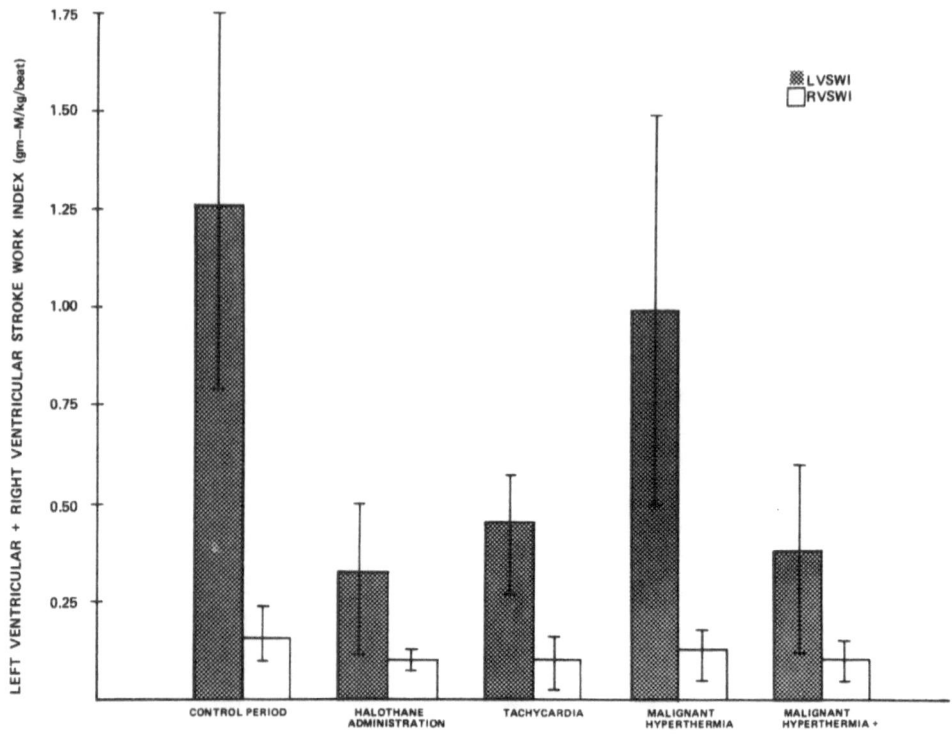

FIGURE 4.8. Cardiac work indices during MH development.

tients develop MH without muscle rigor.[19] We had two pigs in this series that developed MH without muscle rigor. Therefore, we are forced to conclude that muscle rigor per se is not the determining factor in the development of MH but is a late factor that reflects a series of biochemical derrangements that allow a sufficient intracellular influx of calcium ions (up to $10^{-5}M$) or higher and thereby initiates the muscle contractile mechanisms.[20] However, the development of a high temperature without muscle rigor suggests that heat production in skeletal muscle can proceed at a high rate (probably via a membrane ion pumping mechanism or a substrate cycle) over a period of time (usually 30 minutes) with a slow temperature rise. The temperature rise has all of the biochemical components of increased metabolic rate, such as lactate, increased O_2 consumption, fatty acid mobilization, increased CO_2 production, decreased creatinine phosphate, increased adenosine triphosphate (ATP) turnover, and activation of other metabolic processes.

Our recent success in using diltiazem, a calcium channel blocker, suggests that calcium channel activation,[21] presumably by catecholamines

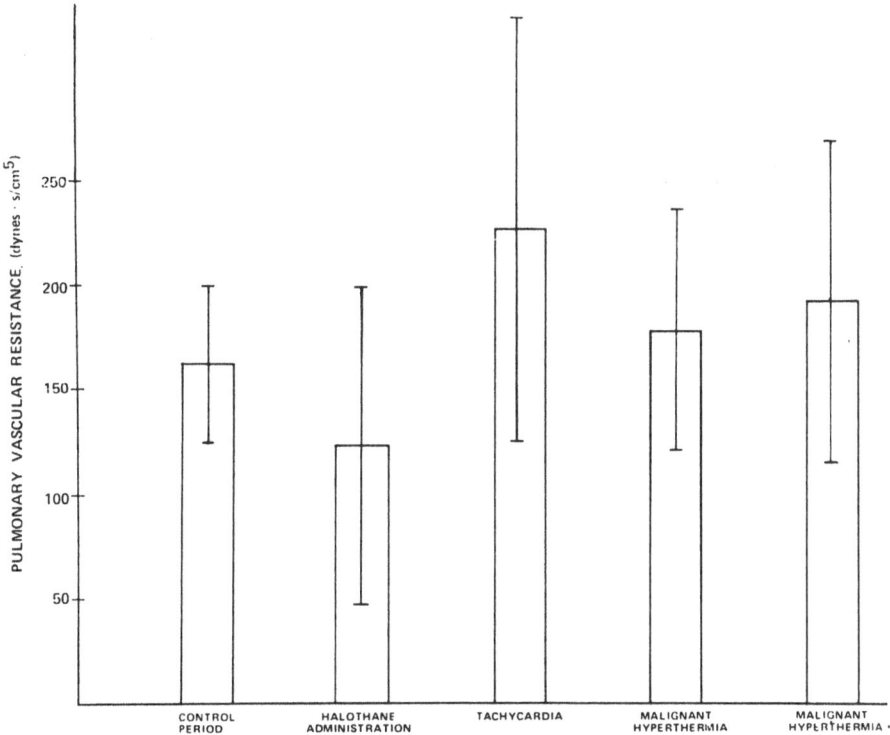

FIGURE 4.9. Pulmonary vascular resistance during the development of MH.

(primarily norepinephrine), would be the key step in activating the heat-producing metabolic processes.

Summary

Twelve malignant hyperthermia susceptible (MHS) pigs were anesthetized with thiopental (22 mg/kg), intubated, and mechanically ventilated, and invasive lines were installed for monitoring electrocardiogram (ECG), heart rate (HR), arterial pressure (AP), pulmonary artery pressure (PAP), pulmonary capillary wedge pressure (PCWP), central venous pressure (CVP), venous O_2 saturation, arterial O_2 saturation, end tidal CO_2, cardiac output (CO), and core and rectal temperature. MH was induced with 2% halothane. Heart rate increased significantly ($P < 0.05$) 5 minutes before any other recorded symptom of impending MH. Total peripheral resistance decreased from 1625 to 762 dynes·sec/cm^5 as the MH syndrome developed. Pulmonary vascular resistance increased from 164 to 225 dynes·sec/cm^5.

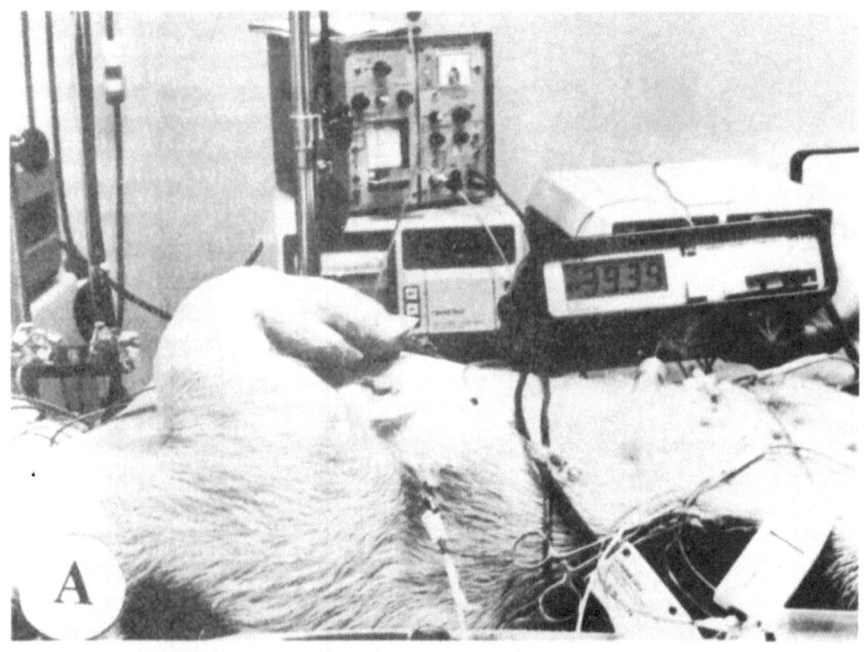

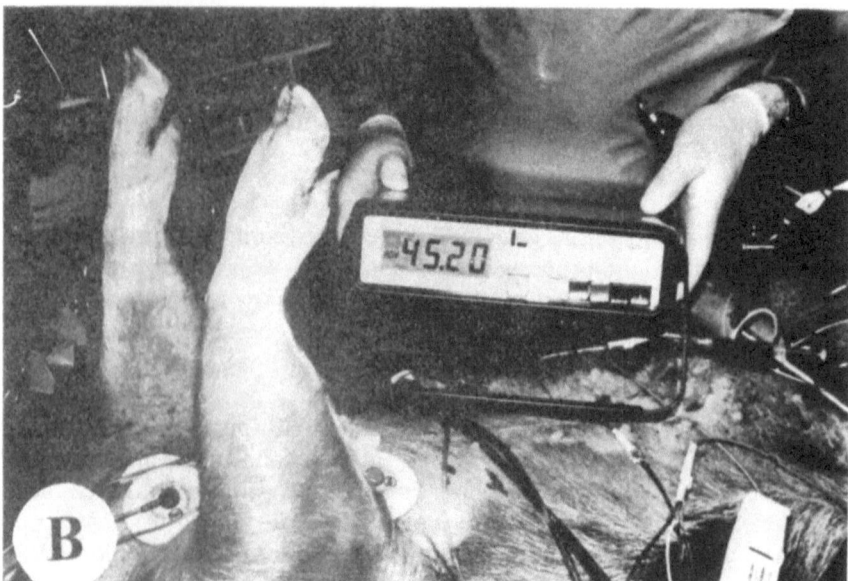

FIGURE 4.10. The development of a core temperature up to 45.20°C in an MHS pig. A. Control period, core temperature 39.39°C. B. MH with rigor, core temperature 45.20°C.

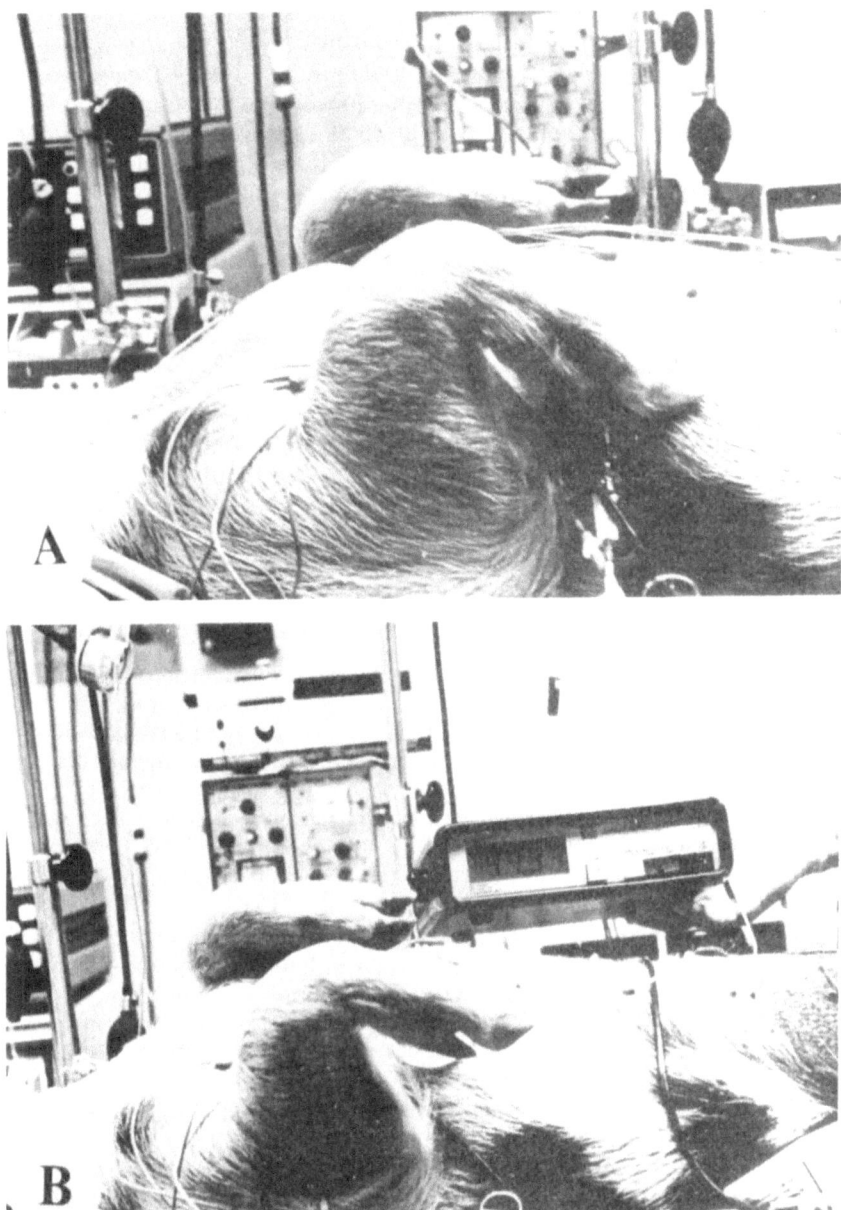

FIGURE 4.11. The development of a core temperature up to 44.32°C in MHS pig without muscle rigor. A. Control period. B. MH without rigor, core temperature 44.31°C.

Pulmonary edema occurred in two pigs. Left ventricular stroke work index dropped from 1.25 to 0.32 g·m/kg/beat during MH. Venous O_2 saturation decreased at a rate of 9%/min. Pulmonary artery core temperature increased at a rate of 0.09°C/min. Norepinephrine levels increased (at tachycardia) to 20.79 ng/ml and peaked at 40.58 ng/ml. Two pigs developed MH without muscle rigor. Tachycardia was the first external physiologic symptom of impending MH. Outcross MH pigs exhibited a decreased peripheral resistance in contrast to our inbred animals, which exhibited an increased peripheral resistance. The high norepinephrine levels are probably activating and opening calcium channels as part of the hormonal stimulus for heat production in muscle.

References

1. Williams CH, Stubbs DH, Payne CG, Benedict JD (1976) Role of hypertension in fulminant hyperthermia-stress syndrome. Br Med J 1:628
2. Williams CH, Shanklin MD, Hedrick HB, Muhrer ME, Stubbs DH, Krause GF, Payne CG, Benedict J.D., Hutcheson DP, Lasley JF (1978) The fulminant hyperthermia-stress syndrome: Genetic aspects, hemodynamic and metabolic measurements in susceptible and normal pigs, in Aldrete JA, Britt BA (eds) *Proceedings 2nd International Symposium on Malignant Hyperthermia*. New York: Grune & Stratton, pp 113–140
3. Williams CH (1976) Some observations on the etiology of the fulminant hyperthermia-stress syndrome. Perspect Biol Med 20(1):120–130
4. Williams CH, Houchins C, Shanklin MD (1975) Energy metabolism in pigs susceptible to the fulminant hyperthermia stress syndrome. Br Med J 3:411–413 and Br Med J 1:724 (1977) errata
5. Britt BA, Gordon RA (1969) Three cases of malignant hyperthermia with special consideration of management. Can Anaesth Soc J 16(2):99–105
6. Hall GM, Lucke JN, Lovell R, Lister D (1980) Porcine malignant hyperthermia. VII. Hepatic metabolism. Br J Anaesth 52:165–171
7. Hall GM, Lucke JN, Lister D (1977) Porcine malignant hyperthermia. V. Fatal hyperthermia in the Pietrain pig associated with the infusion of α–adrenergic agonists. Anaesthesia 49:855–863
8. Gronert GA, Milde JH, Theye RA (1977) Role of sympathetic activity in porcine malignant hyperthermia. Anesthesiology 47:411–415
9. Gronert GA, Heffron JJA, Milde JH, Theye RA (1977) Porcine malignant hyperthermia: Role of skeletal muscle in increased oxygen consumption. Can Anaesth Soc J 24:103–109
10. Lucke JN, Hall GM, Lister D (1977) Anaesthesia of pigs sensitive to malignant hyperthermia. Vet Rec 100:45–48
11. Hall GM, Bendall R. Lucke JN, Lister D (1976) Porcine malignant hyperthermia. II. Heat production. Br J Anaesth 48:305–308
12. Hall GM, Lucke JN, Lister D (1976) Porcine malignant hyperthermia. IV. Neuromuscular blockade. Br J Anaesth 48:1135–1141
13. Jones EW, Nelson TE, Anderson IL, Kerr DD, Burnap TK (1972) Malignant hyperthermia of swine. Anesthesiology 36:42–51
14. Hall LW, Trim CM, Woolf N (1972) Further studies of porcine malignant hyperthermia. Br Med J 2:145–148

15. Berman MC, Harrison GG, Bull AB, Kench JE (1970) Changes underlying halothane-induced malignant hyperpyrexia in Landrace pigs. Nature 225:653–655

16. Wingard DW, Gatz EE (1978) Some observations on stress–susceptible patients, in Aldrete JA, Britt BA (eds) *Proceedings. 2nd International Symposium on Malignant Hyperthermia, Denver, April 1977*. New York: Grune & Stratton, pp 363–372

17. Wingard DW (1974) Malignant hyperthermia–Acute stress syndrome of man? Lancet 2:1450

18. Hoech GP Jr, Roberts JT, Williams CH, Waldman SD, Simpson ST, Trim C, Brazile J (1980) Prevention of porcine malignant hyperthermia with metocurine, in Lomax P, Cox B, Milton AS, Schonbaum E (eds) *Thermoregulatory Mechanisms and Their Therapeutic Implications. 4th International Symposium Pharmacology of Thermoregulation, Oxford, 1979*. Basel: Karger, pp 137–141

19. Britt BA, Kalow W (1968) Hyperrigidity and hyperthermia associated with anesthesia. Ann NY Acad Sci 151:947–958

20. Muir, WW (1982) Effects of calcium inhibitory compounds upon the cardiovascular system, in Rahwan RB, Witiak DT (eds) *Calcium Regulation by Calcium Antagonists*, ACS Symp Series 201. Washington DC: American Chemical Society, pp 39–71

21. Williams CH, Dozier SE, Ilias WK, Fulfer RT, Zukaitis MG, Hoech GP Jr (1985) Treatment of malignant hyperthermia (MH) with diltiazem. Fed Proc 44(5):1638

CHAPTER 5

Plasma Levels of T_4, T_3 and rT_3 During Malignant Hyperthermia

R.R. Anderson, M.A. Akasha, D.A. Nixon, and Charles H. Williams

Introduction

Malignant hyperthermia (MH) is a malady affecting humans[1] as well as pigs,[2] dogs,[3] and horses.[4] The condition may be brought on by certain types of stressors including high ambient temperature or fright, and also by halothane anesthesia.[5] The latter has been used to a considerable extent in triggering MH in certain breeds and families of domestic swine.[6-8]

Sudden rises in body temperature are thought to be mediated by neurohormones and neurotransmitters with a rapid mechanism of action.[9] However, the control of basal metabolic rate by more general acting hormonal modulators, such as thyroid hormones, may also play a significant role. Several years ago, Eighmy et al.[10] reported circulating thyroxine (T_4) levels of MH-positive pigs as significantly higher than those of normal controls. A challenge with halothane anesthesia to trigger the MH syndrome resulted in elevated T_4 plasma levels in both MH and control pigs, but the MH pigs had consistently and significantly higher plasma T_4 levels than the controls.

Improvements in radioimmunoassay (RIA) in recent years have enabled the researcher to measure plasma thyroid hormone concentrations. In addition to T_4 and triiodothyronine (T_3), the inactive metabolite of T_4 known as reverse triiodothyronine (rT_3) may now be measured routinely by RIA. The physiologic significance of plasma rT_3 concentration is that it may reflect the cellular conversion of T_4 into the inactive rT_3 in contrast to the cellular conversion of T_4 to the more active T_3.[11] Such a shift in thyroid hormone conversion at the cellular level should occur under conditions in which further metabolic stimulation of the cell by T_3 is unwarranted, and the control to prevent overstimulation of metabolism is to form the calorigenically inactive form of T_3, namely, rT_3. Normal circulating levels of rT_3 in intact pigs have been estimated to be 490 pg/ml of serum.[12]

The present study was initiated in pigs to determine if thyroid hormones reflected a physiologic mechanism that might aid in explaining the cause for MH. The data show a relationship between the intensity of the MH syndrome and a time–concentration response in the conversion of T_4 to rT_3 concomitant with a reduction in conversion of T_4 to T_3.

Materials and Methods

The experiments were conducted on MH-susceptible (MHS) pigs that had been screened for susceptibility to the MH syndrome using halothane anesthesia.[9] In the initial experiments, blood samples were obtained from either normal or MHS pigs by venipuncture of the jugular vein at or near the juncture with the anterior vena cava. Heparin was used as the anticoagulant. Plasma samples were frozen on dry ice and stored until assayed for T_4, T_3, or rT_3 in separate assays. RIA kits were obtained from Serono Diagnostics (Braintree, MA) for analyses of plasma T_4, T_3, and rT_3. Volumes of 100 μL plasma were used in the reaction tubes, with samples run in duplicate in all cases. Sensitivities of the assays were 200 pg per tube for T_4 and 12.5 pg per tube for T_3 and rT_3. Intraassay and interassay variations were within 10%. Two trials were run with 20 pigs (10 MHS and 10 normal) in the first and 24 pigs in the second trial. Mean values were subjected to analysis of variance and Student's t-test or Fisher's multiple range test.[13]

In the second series of experiments, eight MHS pigs were premedicated with thiopental (22 mg/kg) via an ear vein. Sixty percent nitrous oxide and thiopental drip were used for surgical cutdowns. A cannula was inserted in the jugular vein and carotid artery. An Opticath catheter was inserted in the external iliac vein and advanced to the wedge position in the pulmonary artery. Invasive monitoring consisted of arterial pressure, pulmonary artery pressure, pulmonary artery wedge pressure, central venous pressure, cardiac output via thermodilution, core temperature, rectal temperature via rectal thermistor, venous oxygen saturation, heart rate, and dP/dT via a Millar catheter inserted into the left ventricle via the carotid artery. After a 30-minute recovery and control period, a control blood sample was obtained just prior to administration of 2% to 3% sevoflurane. The sevoflurane anesthetic was continuously monitored, and a blood sample was taken every 30 minutes until the animal showed clear signs of the MH syndrome. Sampling was usually terminated after 120 minutes but did extend to 330 minutes in Pig 4. As in the first experiment, all plasma samples were analyzed for T_4, T_3, and rT_3 in duplicate. Regression analyses for T_4, T_3, and rT_3 were conducted to determine whether or not a significant regression existed for rises or falls in the parameters over time ($P < 0.05$). Computerized programs were used for this purpose.

R.R. Anderson et al.

TABLE 5.1. Thyroid hormone concentrations in plasma of malignant hyperthermia-susceptible and normal pigs: Trial 1

Animal	Normal ng/ml			MHS ng/ml		
	T_4	T_3	rT_3	T_4	T_3	rT_3
1	28.4	1.12	0.53	30.1	0.69	0.81
2	22.0	0.49	0.38	34.0	0.52	1.25
3	38.7	0.83	0.58	32.5	0.69	1.35
4	19.5	0.79	0.45	33.2	0.44	0.74
5	20.6	0.63	0.50	24.1	0.51	0.64
6	14.1	0.61	0.48	29.2	0.72	0.71
7	20.9	0.64	0.63	21.6	1.23	0.70
8	17.3	1.07	0.46	32.9	1.00	0.97
9	23.4	1.20	1.30	35.8	1.16	0.89
10	13.4	0.92	0.41	27.1	1.14	0.80
Mean	21.8	0.83	0.56	30.1	0.81	0.89

Results

Two trials of the MHS pigs versus normal pigs resulted in T_4, T_3, and rT_3 plasma values shown in Tables 5.1 and 5.2. In Trial 1 (Table 5.1) the plasma T_4 of 10 pigs previously designated as MHS was 21.8 ng/ml, wheras in normal pigs, the T_4 averaged 30.1 ng/ml. Because the variation was great, no significant differences were found. The means of plasma T_3 were

TABLE 5.2. Thyroid hormone concentrations in plasma of malignant hyperthermia-susceptible and normal pigs: Trial 2

Animal	MHS ng/ml			Normal ng/ml		
	T_4	T_3	rT_3	T_4	T_3	rT_3
1	18.8	0.94	0.71	18.7	0.63	0.29
2	13.0	0.86	0.68	20.3	0.53	0.29
3	15.7	1.13	0.84	28.7	0.97	0.46
4	10.0	1.33	0.48	16.4	0.77	0.38
5	18.3	0.67	1.42	21.2	0.57	0.43
6	14.9	0.51	0.54	21.9	0.74	0.38
7	16.3	0.73	1.00	12.8	0.56	0.30
8	15.9	0.79	0.60	23.5	0.98	0.64
9	22.2	0.41	0.63	16.2	0.96	0.65
10	21.1	1.20	0.80	38.0	0.80	0.58
11	20.7	0.58	0.96	28.8	0.78	0.97
12	21.1	0.58	0.73	18.4	0.78	1.32
Mean	17.3	0.81	0.78	22.1	0.76	0.56

0.83 ng/ml for MHS pigs and 0.81 ng/ml for normal pigs. Obviously, these means were not statistically different. Mean rT_3 plasma concentrations were 0.56 ng/ml in MHS pigs and 0.89 ng/ml in normal pigs. Although the difference appears to be great, no significant differences were found in the data. In Trial 2 (Table 5.2), the trends were similar in that the mean plasma T_4 concentration in 12 MHS pigs was lower than in 12 normal pigs, but the difference was not significant. Plasma T_3 levels were similar in both groups. Concentrations of rT_3 were fairly large but were in the opposite direction from Trial 1. Again, no statistically significant differences were found.

In the second experiment, pigs identified previously as MHS were sampled under controlled conditions in the operating room. The control samples were taken when the animals were subjected to thiopental and nitrous oxide and prior to the halogenated anesthetic. Control values of thyroid hormones in plasma were 16.9 ng/ml for T_4, 0.60 ng/ml for T_3, and 1.065 ng/ml for rT_3. The mean concentration of T_3 in the eight pigs in this experiment was similar to those in Trials 1 and 2. However, the concentration of T_4 was lower and that of rT_3 was higher than the levels of plasma hormones found in the first experiment. These results suggest an increased conversion of T_4 to rT_3 in the MHS pigs subjected to thiopental/nitrous oxide anesthesia and surgical cutdowns.

As the effects of the halogenated anesthetic were manifested over time, a trend was noted in which T_4 tended to decrease from 17.9 ng/ml at 30 minutes to 15.8 ng/ml at 120 minutes into the syndrome. There was a drop in T_3 plasma concentrations as well. The T_3 started at 0.60 ng/ml, and it was reduced continuously to 0.43 ng/ml by 90 minutes into the experiment with the halogenated anesthetic. As the T_3 was reduced, the rT_3 tended to increase at an even faster rate. Beginning at zero time, the rT_3 increased gradually from 1.065 ng/ml to 1.474 ng/ml at 120 minutes into the syndrome.

TABLE 5.3. Thyroid hormone levels in plasma samples of 8 malignant hyperthermia-positive pigs in response to sevoflurane anesthesia over time

Time in Relation to Invitation of Sevoflurane	T_4 ng/ml	T_3 ng/ml	rT_3 ng/ml
Control	16.9	0.60	1.065
Thiopental N_2O anesthesia			
30 minutes	17.9	0.57	1.130
60 minutes	16.4	0.50	1.215
90 minutes	17.3	0.43	1.371
120 minutes	15.8	0.47	1.474
150 minutes	16.3	0.43	1.288
180 minutes	15.3	0.46	1.502

A regression analysis was performed on each set of data for the three thyroid hormones. Although T_4 did not show a statistically significant regression ($P > 0.05$), the downward regression for T_3 and the upward regression for rT_3 were statistically significant ($P < 0.05$). These data are presented in tabular form (Table 5.3).

Each animal experienced a particularly individualized response in the changes of thyroid hormones during the MH syndrome in response to the halogenated anesthetic. Pig 3 (Figure 5.1) had a response very similar to the pattern for mean values, whereas Pig 4 (Figure 5.2) showed a slow response to the increase in plasma rT_3. Pig 8 had a fairly quick response by increasing rT_3 within the first 60 minutes (Figure 5.3). Pig 9 showed little change in T_3 but a sharp rise in rT_3 from 60 minutes to 120 minutes (Figure 5.4). Pig 10 experienced an early drop in T_3 between 0 and 60 minutes, with a delayed increase in rT_3 between 90 and 120 minutes (Figure 5.5). Pig 15 showed an initial spike in T_3, reminiscent of our earlier findings. However, the T_3 concentration decreased rapidly from 30 to 60 minutes in this animal. The rT_3 rose from the 60-minute sample to the 90-minute sample and still higher at the 120-minute sample (Figure 5.6). Pig 16 began at zero time with a very high rT_3 level. This increased 15% between 90 and 120 minutes; T_3 showed a sawtooth pattern in this pig (Figure 5.7).

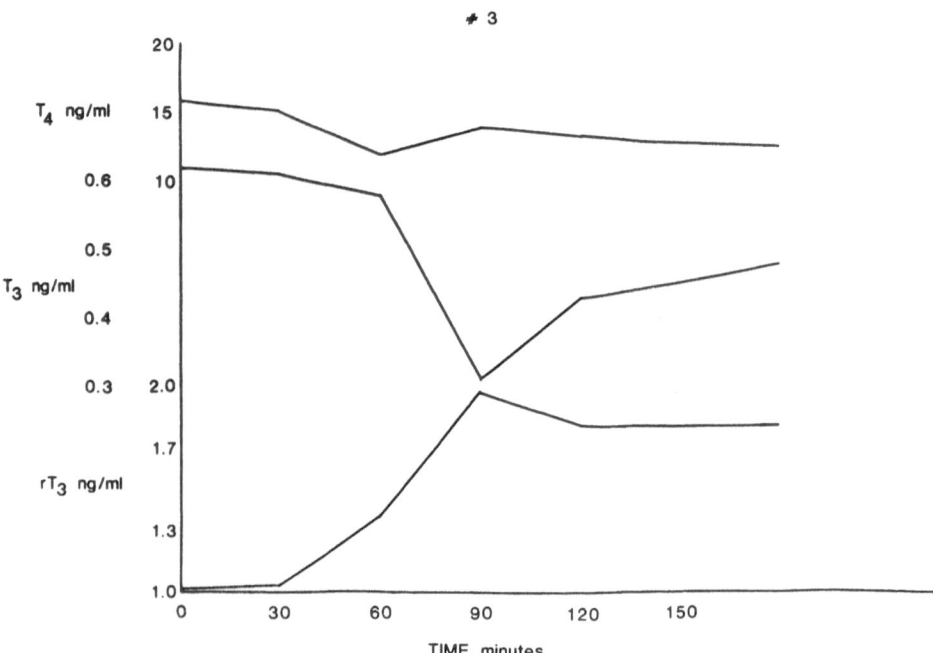

FIGURE 5.1. Concentrations of T_4, T_3, and rT_3 in serum samples of Pig 3 after exposure to sevoflurane anesthesia.

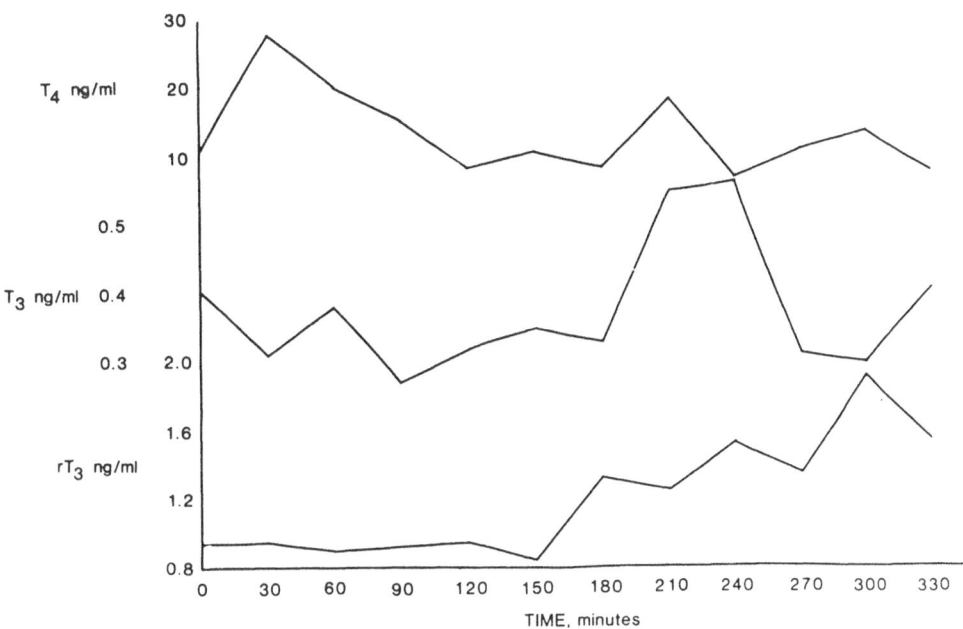

FIGURE 5.2. Concentrations of T_4, T_3, and rT_3 in serum samples of Pig 4 after exposure to sevoflurane anesthesia.

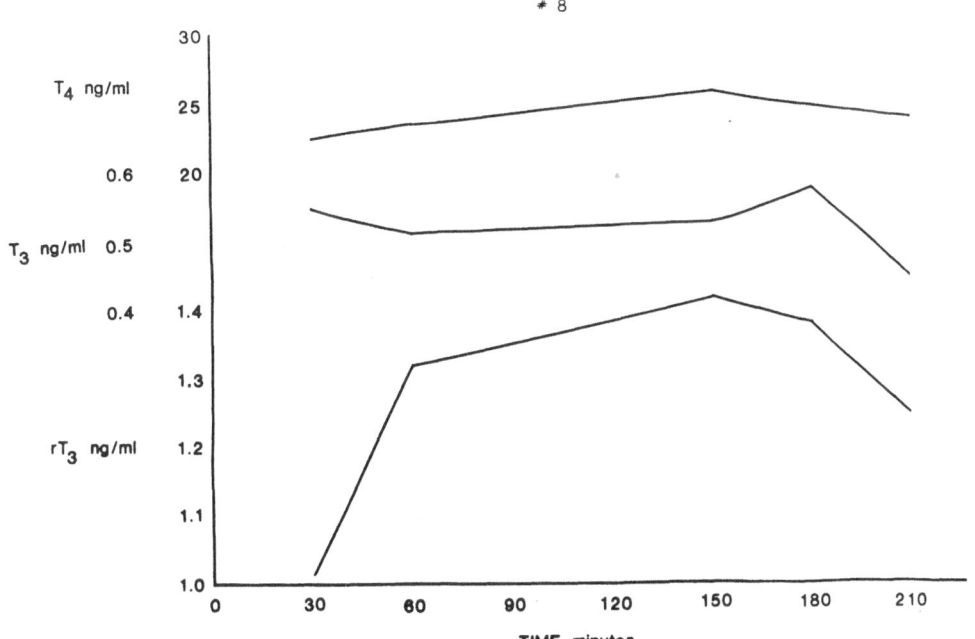

FIGURE 5.3. Concentrations of T_4, T_3, and rT_3 in serum samples of Pig 8 after exposure to sevoflurane anesthesia.

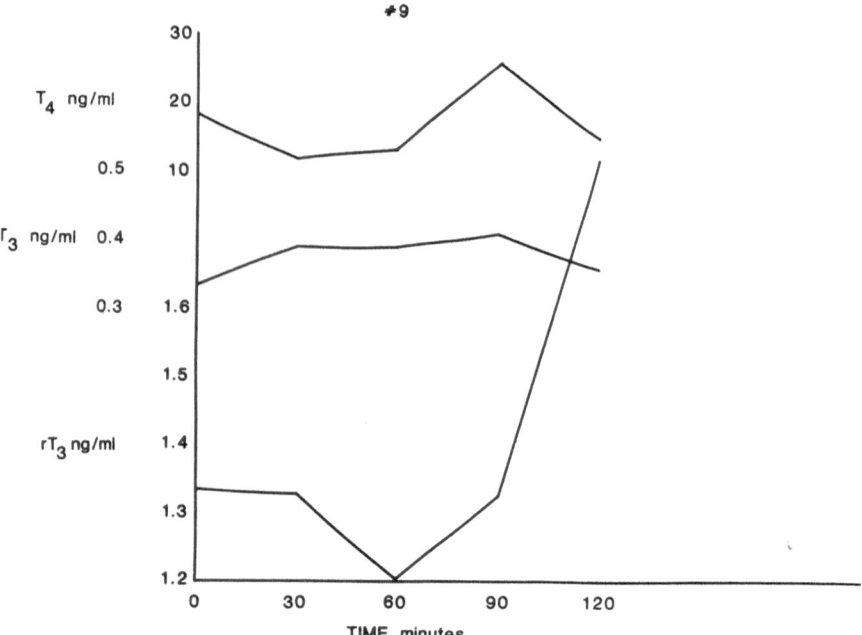

FIGURE 5.4. Concentrations of T₄, T₃, and rT₃ in serum samples of Pig 9 after exposure to sevoflurane anesthesia.

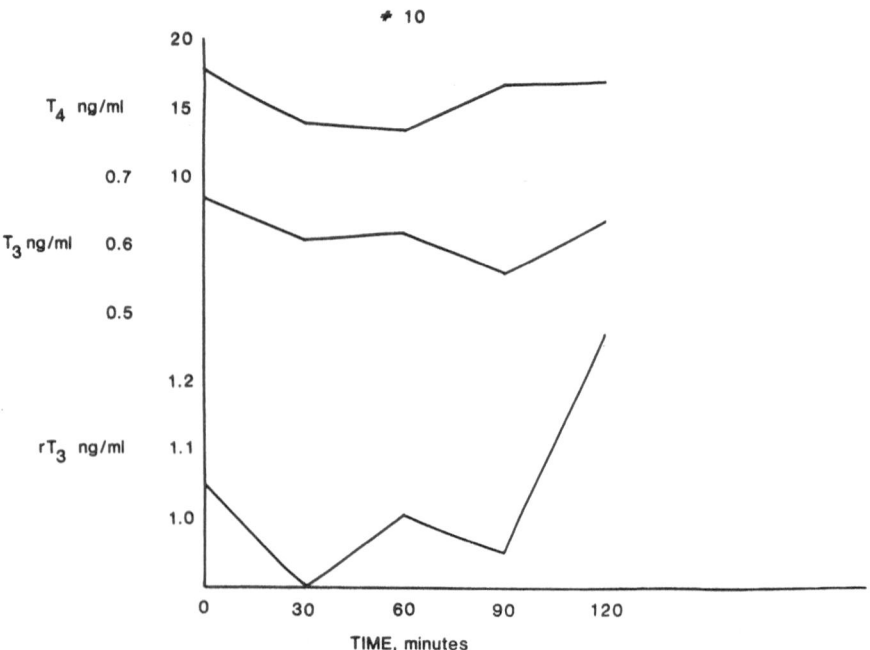

FIGURE 5.5. Concentrations of T₄, T₃, and rT₃ in serum samples of Pig 10 after exposure to sevoflurane anesthesia.

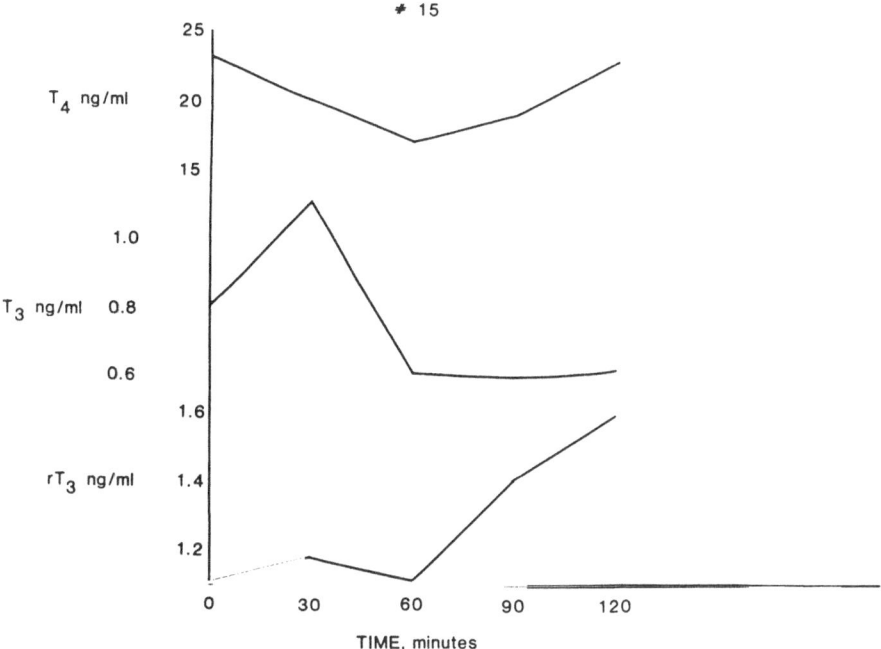

FIGURE 5.6. Concentrations of T_4, T_3, and rT_3 in serum samples of Pig 15 after exposure to sevoflurane anesthesia.

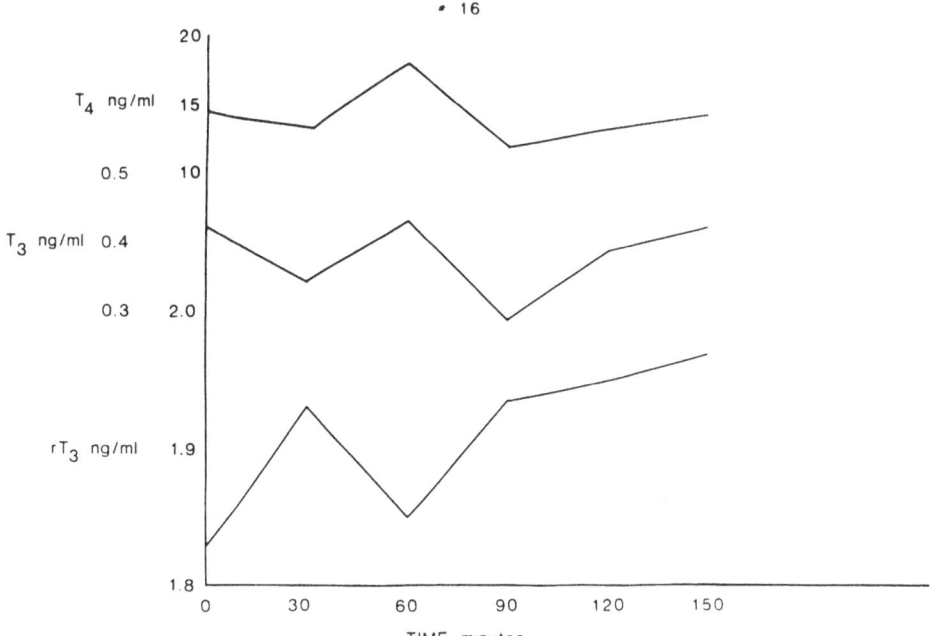

FIGURE 5.7. Concentrations of T_4, T_3, and rT_3 in serum samples of Pig 16 after exposure to sevoflurane anesthesia.

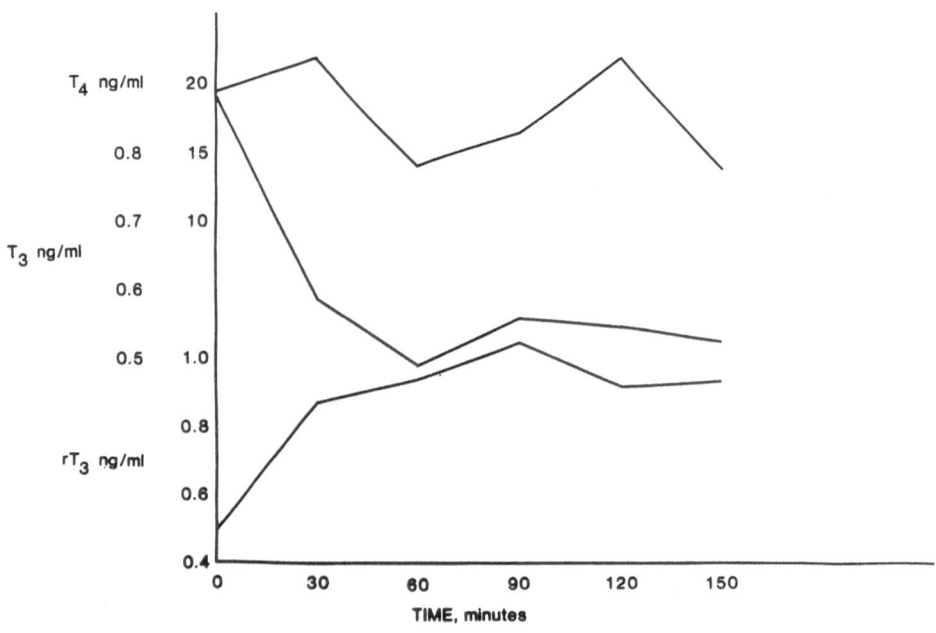

FIGURE 5.8. Concentrations of T_4, T_3, and rT_3 in serum samples of Pig 17 after exposure to sevoflurane anesthesia.

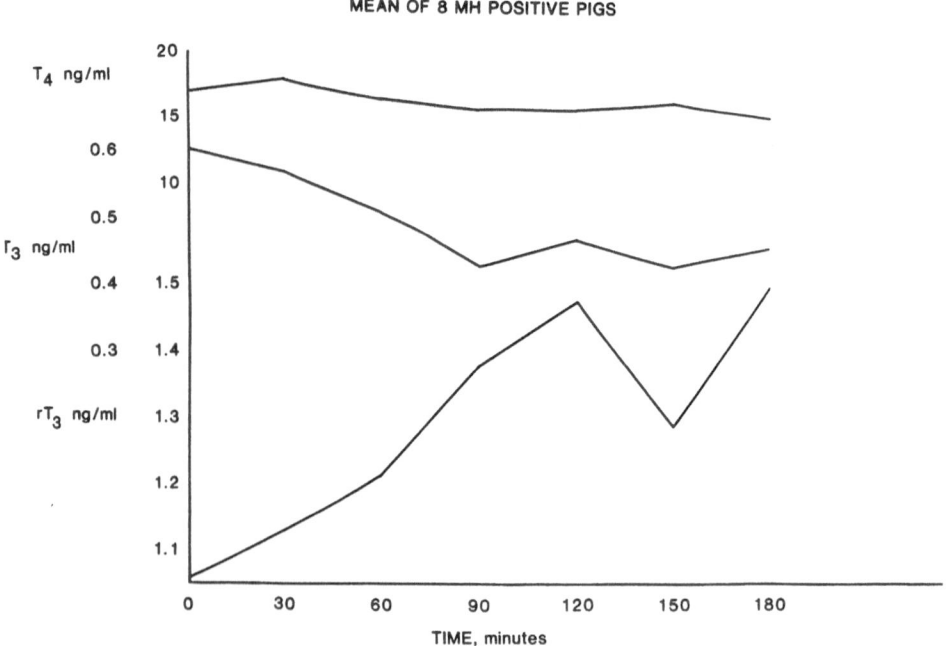

FIGURE 5.9. Composite of mean concentration of T_4, T_3, and rT_3 in serum samples of 8 MHS pigs after exposure to sevoflurane anesthesia.

Pig 17 experienced a rapid drop in plasma T$_3$, reaching a low at 60 minutes and maintaining a low level to 150 minutes. The rT$_3$ in this pig rose gradually between 0 and 90 minutes (Figure 5.8). A composite of the mean values obtained from 8 MHS pigs is shown in Figure 5.9. The decrease in plasma T$_3$ and a concomitant increase in plasma rT$_3$ represent significant regression lines. The calculated decrease in T$_3$ was 1.9 ng/ml per minute, and the increase in rT$_3$ was 3.0 ng/ml per minute.

Discussion

Thyroid hormones are well known as regulators of metabolic processes in most body tissues. Since hyperthyroidism is defined by many investigators as a rise in oxygen consumption by body tissues at a level at least 30% above normal, the thyroid hormones are logical contenders as factors to be considered in the MH syndrome. Hyperthyroidism in MHS pigs was demonstrated by Eikelenboom and Weiss[14] with Pietrain pigs. Hall and Lister[15] suggested that the utilization rates of T$_4$ and T$_3$ were increased in stress-susceptible Pietrain pigs. A fall in the serum-free T$_4$ index of Pietrain pigs subjected to the MHS in response to succinylcholine suggested a rapid utilization of T$_4$. When large doses of T$_3$ were administered, the death of pigs experiencing the syndrome was prevented.[16] Pigs selected for stress resistance were thyroidectomized and then made hyperthyroid with large doses of T$_4$. After slaughter, the muscles of these pigs had significantly higher concentrations of lactic acid than had the controls.[17] Pigs experiencing the MH syndrome are also found to have high levels of muscle lactic acid and pale, soft, exudative pork.[18,19] Previous studies comparing MHS versus normal pigs showed plasma T$_4$ and T$_3$ levels to be elevated in the MH pigs whether subjected to halothane anesthesia or not anesthetized at all.[10] These findings suggested a positive role for T$_4$ and T$_3$ in potentiating the fulminating actions of norepinephrine in the MH syndrome. The significance of this catecholamine in bringing about vasoconstriction at the periphery to cause hypertension has been demonstrated.[20,22]

Hyperthyroidism has been shown to accelerate substrate cycling.[23] The rate of the glucokinase-glucose-6-phosphatase substrate cycle and the phosphofructokinase-fructose biphosphate cycle increased significantly above normal in hyperthyroid rats. The phosphofructokinase-fructose biphosphate cycle was increased in activity in muscle of MHS pigs.[24]

The evidence of reduced plasma T$_3$ in the presence of increased plasma rT$_3$ probably reflects a switch in the control mechanisms for production of these two triiodothyronines from T$_4$. Although a considerable amount of T$_3$ used by peripheral tissues, such as muscle and liver, may originate from the thyroid gland, much (estimated to be at least 50%) is derived from T$_4$ as a result of intracellular conversion.[11] The primary source of circulating rT$_3$ is from the metabolism of T$_4$ at the peripheral tissue level.[25]

Conversion of T_4 generally is in the preferred direction of T_3 production.[26] This requires the activity of an enzyme known as 5′–monodeiodinase, or outer ring deiodinase. However, certain circumstances and tissues may preferentially convert T_4 to rT_3 in the presence of an active form of 5–monodeiodinase or inner ring deiodinase. The rat placenta is a prime example of this activity.[27] The present study demonstrates a reduction in T_3 and an increase in rT_3 during the course of the MH syndrome triggered by a halogenated anesthetic. This suggests a change in activities of intracellular enzymes. It would appear that inner ring deiodinase activity increases whereas outer ring deiodinase activity decreases as the MH syndrome intensifies. This generally is accompanied by a rise in body temperature. Since conditions of extreme hyperthermia are not commonly experienced by peripheral cells, such as muscle, liver, and kidney, the conversion of T_4 to rT_3 in these tissues is not readily demonstrable. The present findings suggest that because rT_3 is much less active than T_3 in stimulating metabolic processes, this step is a final attempt to preserve the life of the organism by decreasing metabolism and heat load.

Summary

Pigs were triggered for the MH syndrome by using sevoflurane in the anesthesia regimen. Under nonanesthetized normal active conditions, blood samples were obtained from previously implanted jugular catheters. The plasma thus obtained was analyzed by RIA for thyroxine (T_4), triiodothyronine (T_3) and reverse triiodothyronine (rT_3). Although normal pigs had slightly higher plasma T_4 concentrations and lower T_3 concentrations, with no apparent differences in rT_3 levels, the differences were not significant on statistical evaluation. Further study by time sequence sampling MH-positive pigs during sevoflurane anesthesia challenge over a period of up to 4 hours revealed a pattern of significant changes in plasma T_3 and rT_3 concentrations. As the MH symptoms increased in intensity, the T_3 concentrations decreased and the rT_3 concentrations increased. These findings were interpreted as a reflection of cellular changes from the normally occurring conversion of T_4 to T_3 to the less calorigenic pathway of T_3 conversion to rT_3. Thus, a mechanism whereby the MH-positive pig attempts to counteract the rapid rise in body temperature has become apparent.

Acknowledgements. This study is a contribution from the Missouri Agricultural Experiment Station Journal Series Number 9969. Funding for RIA kits was from the People's Committee of the Libyan Government. The MHS pig colony was maintained by a grant from Anesthesia Associates of Kansas City, Missouri. The experimental surgery facilities, supplies, and equipment were supported by a seed grant from Texas Tech School of Medicine, Organon Inc. Grant NC 45-007-83-0019 and Bio-dynamics Inc. Grant 84-2900.

References

1. Britt BA (1974) Malignant hyperthermia: A pharmacogenetic disease of skeletal and cardiac muscle. N Engl J Med 290:1140–1150
2. Nelson TE, Jones EW, Henrickson RL, Falls SN, Kerr DD (1974) Porcine malignant hyperthermia: Observations on the occurrence of pale, soft, exudative musculature among susceptible pigs. Am J Vet Res 35:349–350
3. O'Brien PJ, Forsyth GW, Olexson DW, Thatte HS, Addis PB (1984) Canine malignant hyperthermia susceptibility: Erythrocytic defects: Osmotic fragility, glucose-6-phosphate dehydrogenase deficiency and abnormal calcium homeostatis. Can J Comp Med 48:381–389.
4. Hildrebrand SV, Howitt GA (1983) Succinylcholine infusion associated with hyperthermia in ponies anesthetized with halothane. Am J Vet Res 44:2280–2284
5. Britt BA (1982) Malignant hyperthermia—A review, in Milton A (ed) *Handbook of Experimental Pharmacology, Pyretics and Antipyretics*. Heidelberg: Springer-Verlag, pp 547–615
6. Ollivier L, Sellier P, Monin G (1975) Genetic determination of the malignant hyperthermia syndrome in the Pietrain pig. Ann Genet Sel Anim 7:159–166
7. Gregory NG, Lister D (1981) Autonomic responsiveness in stress-sensitive and stress-resistance pigs. J Vet Pharm Ther 4:67–76.
8. Andresen E, Jensen P, Barton-Gade P (1981) The porcine Hal locus: A major locus exhibiting overdominance. Z Tierz Zuechtungsbiol 98:170–175
9. Williams CH (1976) Some observations on the etiology of the fulminant hyperthermia-stress syndrome. Perspect Biol Med 20:120–130
10. Eighmy JJ, Williams CH, Anderson RR (1978) The fulminant hyperthermia-stress syndrome: Plasma thyroxine and triiodothyronine levels in susceptible and normal pigs and man, in Aldrete JA, Britt BA (eds) *Malignant Hyperthermia*. New York: Grune & Stratton, pp 161–173
11. Chopra IJ, Solomon DH, Chopra U, WU SY, Fisher DA, Nakamura J (1978) Pathways of metabolism of thyroid hormones. Recent Prog Horm Res 34:521–567
12. Rudas P (1980) Reverse-triiodothyronine (rT$_3$) in domesticated animals. Acta Physiol Acad Sci Hung 56:74
13. Snedecor GW, Cochran WG (1967) *Statistical Methods*, 6th ed. Ames, Iowa: Iowa State University Press
14. Eikelenboom G, Weiss GM (1972) Breed and exercise influence on T$_4$ and PSE. J Anim Sci 35:1096
15. Hall GM, Lister D (1975) Endocrine aspects of malignant hyperthermia in Pietrain pigs, in Arios A, Llaurado R, Nalda MA, Lunn JN (eds) *Recent Progress in Anesthesiology and Resuscitation*. New York: American Elsevier, pp 357–358
16. Lister D (1973) Correction of adverse response to suxamethonium of susceptible pigs. Bri Med J, 1:208–210

17. Marple DN, Nachreiner RF, McGuire JA, Squires CD (1975) Thyroid function and muscle glycolysis in swine. J Anim Sci 41:799–803
18. Kastenschmidt LL, Briskey EJ, Hoekstra WG (1966) Metabolic intermediates in skeletal muscles with fast and slow rates of post-mortem glycolysis. Nature 212:288–289
19. Harrison GG (1972) Pale, soft, exudative pork, porcine stress syndrome and malignant hyperpyrexia—An identity? J So Af Vet Assoc 43:57–63
20. Williams CH, Stubbs DH, Payne CG, Benedict JD (1976) Role of hypertension in fulminant hyperthermia-stress syndrome. Br Med J 1:628
21. Davis TP, Gehrke CW Jr, Williams CH, Gehrke CW, Gerhardt KO (1982) Precolumn derivatization and high-performance liquid chromatography of biogenic amines in blood of normal and malignant hyperthermic pigs. J Chromatogr 228:113–122
22. Williams CH, Dozier SE, Buzello W, Gehrke CW, Wong JK, Gerhardt KO (1985) Plasma levels of norepinephrine and epinephrine during malignant hyperthermia in susceptible pigs. J Chromatogr 344:71–80
23. Huang MT, Lardy HA (1981) Effects of thyroid states on the Cori cycle, glucose-alanine cycle, and futile cycling of glucose metabolism in rats. Arch Biochem Biophys 209:41–51
24. Clark MG, Williams CH, Pfeifer WF, Bloxham DP, Holland PC, Taylor CA,Lardy HA (1973) Accelerated substrate cycling of fructose-6-phosphate in themuscle of malignant hyperthermic pigs. Nature 245:99–101
25. Kaplan MM, Schimmel M, Utiger RD (1977) Changes in serum 3,3'5–triiodothyronine (reverse T_3) concentrations with altered thyroid hormone secretion and metabolism. J Clin Endocrinol Metab 45:447–455
26. Ferguson DC, Jennings AS (1983) Regulation of conversion of thyroxine to triiodothyronine in perfused rat kidney. Am J Physiol 8:E220–E229
27. Roti E, Fang SL, Braverman LE, Emerson CH (1982) Rat placenta is an active site of inner ring deiodination of thyroxine and 3, 3'5–triiodothyronine. Endocrinology 110:34–37

CHAPTER 6

Malignant Hyperthermia and the Sarcoplasmic Reticulum Membrane: A Review

Mariam A. Marvasti and Charles H. Williams

Introduction

For many years an increased calcium ion activity has been implicated in the development of malignant hyperthermia (MH). Our first investigations of MH in susceptible pigs in 1970 were directed at evaluating the role of Ca^{2+} in MH with studies of isolated skeletal muscle sarcoplasmic reticulum[1-3] and heart, liver, and skeletal muscle mitochondria.[4,5] We came to the conclusion that changes in the Ca^{2+} ion may be involved in MH, but as a secondary event. Numerous studies of sarcoplasmic reticulum fractions from MH susceptible (MHS) pigs and human patients have produced a series of conflicting results from various laboratories, as delineated in this review.

Discussion

The striated muscle cell consists of three membrane systems, all of which are involved in the process of excitation–contraction coupling.[6] The outer surface membrane, called the sarcolemma, surrounds the entire muscle cell and is excited by the innervating nerve to propagate an action potential that initiates excitation–contraction coupling. The invaginations of the sarcolemma into the fiber's interior are called "transverse tubules" or "T tubules." The transverse tubular system is responsible for the spread of depolarization to the interior of the fiber, with an accompanying membrane charge movement.[6]

The muscle cell has an extensive internal membrane system, the sarcoplasmic reticulum, which is a membraneous network that surrounds each myofibril like a fenestrated waterjacket around a cylinder. The sarcoplasmic reticulum is responsible for sequestration of Ca^{2+} to allow relaxation, storage of Ca^{2+} during relaxation, and release of Ca^{2+} to initiate contraction.[6]

In skeletal muscle, the sarcoplasmic reticulum is the sole source of Ca^{2+} during the activation of contraction. The sarcoplasmic reticulum is especially well developed in fast-contracting muscles. The rapid activation–inactivation cycle of the contractile system of these muscles depends on the sudden release of considerable quantities of Ca^{2+}, followed by the complete removal of Ca^{2+} from the sarcoplasm.[6]

The sarcoplasmic reticulum consists of longitudinal reticulum and terminal cisternae. The longitudinal reticulum faces the myofibrils and is referred to as "free sarcoplasmic reticulum" because it does not participate in the formation of any junctions with other membrane systems. The terminal cisternae of the sarcoplasmic reticulum form junctions with the transverse tubules. The combination of a control transverse tubule sandwiched between two terminal cisternae is called a "triad." Triads are located in a regular disposition either opposite the Z line (frog twitch fibers) or opposite the A-I junction (mammalian muscle). The junctional sarcoplasmic reticulum membrane is that portion of the terminal cisternae that is in direct apposition to the transverse tubule membrane.

Freeze–fracture of the longitudinal reticulum or free sarcoplasmic reticulum reveals a cytoplasmic fracture face that consists of closely packed particles, which represent the intramembraneous hydrophobic portion of the $Ca^{2+} + Mg^{2+}$–ATPase. The major function of the longitudinal reticulum appears to be the sequestration of Ca^{2+} to allow relaxation. The longitudinal reticulum does not appear to be a major site of Ca^{2+} storage, nor does it delay the return of Ca^{2+} to the terminal cisternae.[6]

Meissner[7] was the first to purify skeletal sarcoplasmic reticulum by sucrose gradient centrifugation and to show that the isolated sarcoplasmic reticulum consisted of a heterogeneous population of vesicles. He found that sarcoplasmic reticulum vesicles with different buoyant densities (light and heavy vesicles) differed with respect to protein composition and electron–density. He suggested that the heavy sarcoplasmic reticulum vesicles are derived from the terminal cisternae of the sarcoplasmic reticulum and that the light sarcoplasmic reticulum vesicles are derived from the longitudinal sarcoplasmic reticulum.

Sarcoplasmic reticulum vesicles are identified by their ability to accumulate Ca^{2+} from ATP-containing solutions and by their Ca^{2+}–dependent ATPase activity. Ebashi[8] was the first to show that isolated sarcoplasmic reticulum vesicles accumulate Ca^{2+} with high affinity in the presence of ATP.

Hasselbach[9] was the first to describe Ca^{2+} accumulation as active transport against a Ca^{2+} gradient that derives its energy from the hydrolysis of ATP through an ATPase enzyme incorporated in the sarcoplasmic reticulum membrane.

The major protein components of the skeletal sarcoplasmic reticulum vesicles are $Ca^{2+} + Mg^{2+}$–ATPase (105,000 daltons), calsequestrin (63,000 daltons), and the 53,000-dalton glycoprotein.[10] Minor protein components of the sarcoplasmic reticulum vesicles account for 5% to 20% of the total protein. These proteins are possibly very important to sarcoplasmic reticulum function. It is likely that the proteins responsible for excitation–contraction coupling, junctional feet, and Ca^{2+} release channels are all minor protein components of isolated sarcoplasmic reticulum vesicles. $Ca^{2+} + Mg^{2+}$–ATPase consists of a single polypeptide that is able to couple the hydrolysis of one molecule of ATP to the active transport of two Ca^{2+} ions across the sarcoplasmic reticulum membrane.[10]

MacLennan et al.[11] reviewed the role of calsequestrin in Ca^{2+} storage in the sarcoplasmic reticulum. Although the role of the 53,000-dalton glycoprotein in sarcoplasmic reticulum function is not completely understood, it is suggested that it functions in the co–transport of anions during Ca^{2+} transport.

The sarcoplasmic reticulum is primarily responsible for mediating calcium fluxes and thus is intimately involved in regulating contraction and relaxation of muscle. Its role in excitation–contraction coupling has been extensively reviewed over the past decade.

A general model for excitation–contraction coupling in striated muscle involves depolarization of the muscle membrane, resulting in the release of calcium from the sarcoplasmic reticulum to effect contraction by deinhibition of actin–myosin interaction through the binding of calcium to troponin and the consequent conformational changes in the troponin–tropomyosin complex.[12]

Ca^{2+} uptake is most often defined as an ATP–dependent sequestration of Ca^{2+} into the lumen of isolated sarcoplasmic reticulum fractions, usually in the presence of a precipitable anion, such as oxalate or phosphate.[13] Ca^{2+} release from sarcoplasmic reticulum can be initiated by numerous interventions, among them being extravesicular Ca^{2+}, depolarization, ATP, adenosine diphosphate (ADP), inorganic phosphate, alkalinization, and numerous drugs and reagents.

Most comparative studies involving Ca^{2+} release from heavy and light skeletal sarcoplasmic reticulum fractions have involved the use of various drugs (e.g., caffeine, dantrolene sodium, nitrendipine) and ionophores, such as A23187 (calcimycin) and X537A to elicit release.[13]

Methylxanthine, caffeine, caused an increase in force of contraction in both cardiac and skeletal muscles. Caffeine reduced Ca^{2+} uptake by skeletal sarcoplasmic reticulum vesicles and also caused a release of Ca^{2+}, which had been previously sequestered by the sarcoplasmic reticulum.Dantrolene sodium is a potent inhibitor of skeletal muscle contraction and has been used clinically as an antispastic agent. It is generally held that this drug exerts its action by inhibition of Ca^{2+} release by sarcoplasmic reticulum. Concentrations of dantrolene sodium, which decreases contraction in skeletal muscle in vivo and inhibit Ca^{2+} release in skeletal muscle in vitro, have no effect on intact cardiac muscle function or on isolated cardiac sarcoplasmic reticulum.[13]

The dihydropyridine drug, nitrendipine, is a potent calcium channel antagonist that acts as a negative inotropic agent. It binds strongly to cardiac and smooth muscle membranes with dissociation constants in the nano-molar range.[13]

Effects of Local and General Anesthetics on Sarcoplasmic Reticulum Function

Local Anesthesia

Tetracaine, dibucaine, procaine, and lidocaine inhibit calcium transport and calcium pump ATPase activity in both cardiac and skeletal muscle sarcoplasmic reticulum. Low concentrations of these agents can inhibit calcium efflux and caffeine–induced calcium release from sarcoplasmic reticulum vesicles, and higher concentrations increase calcium efflux and inhibit the Ca^{2+}-activated ATPase reaction.[14]

Inhibition of passive calcium efflux from sarcoplasmic reticulum vesicles by low concentrations of local anesthetic agents could be due to their incorporation into the lipid bilayer, to interaction with Ca^{2+} binding sites on the membrane surface, or to direct interaction with the calcium pump ATPase protein itself.[14]

Higher concentrations of local anesthetics accelerate calcium release from calcium–filled sarcoplasmic reticulum vesicles, which may be due to both inhibition of calcium uptake and increased calcium efflux. The inhibition of calcium uptake is associated with inhibition of Ca^{2+} activated ATP hydrolysis, possibly due to a slowing of the rate of decomposition of the phosphorylated intermediates of calcium pump ATPase. The effect of high local anesthetic concentrations could result either from direct interaction with the calcium pump ATPase protein or from modification of phospholipids surrounding this protein within the sarcoplasmic reticulum membrane bilayer.[14]

Maher and Singer[15] reported that treatment of sarcoplasmic reticulum membranes with tetracaine followed by hygroscopic desorption filtration extracted small but significant amounts of membrane components, primarily lipids. These results suggest that local anesthetics, like fatty acids, may substitute for endogenous phospholipids and exert their effects by modifying the fluidity and structure of the bilayer and, hence, the function of the calcium pump ATPase protein.

General Anesthetics

The effects of general anesthetics on sarcoplasmic reticulum function and structure have not been extensively studied, but most studies indicate that these drugs display effects only at concentrations well above those required for clinical anesthesia.

Lain et al.[16] reported that 7 mM halothane decreased calcium uptake of cardiac sarcoplasmic reticulum by 50%. However, clinical concentrations of halothane (1 mM to 2 mM) had no effect on ATP–dependent calcium transport, except at a pH less than 6.3. Both halothane and enflurane (1 mM to 2 mM) have been reported to enhance both calcium and caffeine-induced calcium release from sarcoplasmic reticulum. Heffron and Gronert[17] reported that low halothane concentrations (<0.14 mM) had minimal stimulatory effects on calcium sequestration and no effect on calcium release, whereas higher concentrations (> 0.3 mM) caused moderate inhibition of calcium sequestration and an increase in calcium release. This study suggests a biphasic concentration–dependent action of halothane on calcium sequestration by the sarcoplasmic reticulum that is similar to the effects of local anesthetics and added fatty acids. Halothane has also been observed to inhibit barbiturate binding to dipalmital phosphatidylcholine bilayers. These findings suggest that halothane and possibly barbiturates interact directly with a site on the calcium pump ATPase protein or that they disrupt lipid–protein association within the membrane.

Diethyl ether increases calcium pump ATPase activity and completely abolishes the ability of sarcoplasmic reticulum to accumulate calcium.[18] These findings suggest that both classes of anesthetic agents (general and local) alter sarcoplasmic reticulum function by interacting with the lipid portion of the bilayers, possibly by causing increased fluidization and indirectly altering calcium pump ATPase protein function.

All of these subcellular organelles have been implicated in MH, but the experimental evidence is scant. Many studies suggest that the sarcoplasmic reticulum, which controls intracellular Ca^{2+} fluxes during contraction, is implicated as the defective unit in MH muscle.[19–23]

One study suggested that the rate of calcium binding and rate and capacity of calcium uptake were decreased and spontaneous calcium release was greater in sarcoplasmic reticulum fragments from skeletal muscle of susceptible swine as compared to those from normal swine.[24] However, the effects of halothane did not explain its triggering action in MH because, at clinical concentrations, it stimulated calcium binding by both MH and normal sarcoplasmic reticulum. At higher concentrations, it depressed calcium binding and uptake functions and increased the permeability of both types of sarcoplasmic reticulum.

Ohnishi et al.[25] studied the role of sarcoplasmic reticulum in MH and showed that it demonstrated Ca^{2+} induced calcium release (Ca–ICaR) and halothane–induced Ca^{2+} release (halothane–ICaR). Normal sarcoplasmic reticulum did not demonstrate these release phenomena. Dantrolene inhibited the halothane–ICaR but did not inhibit the Ca–ICaR.

Ruthenium red and tetracaine inhibited both types of Ca^{2+} release. From the measurements of passive Ca^{2+} efflux, it was shown that dantrolene did not affect the Ca^{2+} permeability itself but suppressed only the halothane–induced increment of the permeability. This study showed that halothane disordered the lipid bilayer of MH sarcoplasmic reticulum to a greater extent than it did that of normal sarcoplasmic reticulum. This

halothane-disordering effect on MH sarcoplasmic reticulum was antagonized by dantrolene. Ruthenium red and tetracaine did not antagonize the
halothane disordering effect. It raised the possibility that halothane could
disturb the structure of the lipoprotein complex in MH sarcoplasmic reticulum in such a way that it could open the calcium release channels.
The Ca^{2+} thus released further opens the channels through the Ca-ICaR
mechanism in a positive feedback fashion, thus triggering the MH syndrome. In this study, the efficacy of dantrolene in ameliorating the MH
syndrome was related to the inhibition of this halothane effect.[25]

Nelson and Bee[26] studied the effects of varying temperature on calcium-
binding characteristics of sarcoplasmic reticulum from MHS and a crossbred, less susceptible group (MHX) with control pig muscle and found a
sharp decrease in calcium binding in the MHS and MHX fractions at 35°C
which differentiated the MHS and MHX fractions from controls. They
postulated that these temperature effects on calcium-binding characteristics
of sarcoplasmic reticulum from MHS and MHX muscle may be indicative
of a membrane transition that impairs calcium binding.

In one investigation, Condrescu et al.[27] compared the capacity of calcium
uptake and the $(Ca^{2+} + Mg^{2+})$–ATPase activity of crude sarcoplasmic
reticulum vesicles prepared from MHS and control human muscles. The
calcium uptake was 0.29 ± 0.04 μmol/mg protein/min (mean $\pm$ SEM,
$n = 8$) in sarcoplasmic reticulum vesicles obtained from MHS patients,
whereas in control subjects it was 1.01 ± 0.05 μmol/mg protein/min (mean
$\pm$ SEM, $n = 8$). The $(Ca^{2+} + Mg^{2+})$–ATPase activity was also considerably lower in MH sarcoplasmic reticulum (0.12 ± 0.01 μmol P_i/mg pro-
tein/min, $n = 8$) than in normal (0.36 ± 0.01 μmol P_i/mg protein/min,
$n = 8$). These results were in good agreement with the intracellular free
calcium determination, suggesting that the increase of the free resting calcium concentration in MH muscles might be due to a deficiency in the
capacity of calcium uptake by the sarcoplasmic reticulum. In another
study, Nelson[28] suggested that MH is a consequence of abnormal release
of Ca^{2+} in muscles, which triggers metabolic and contractive events. He
found no significant differences for rate of Ca^{2+} uptake by sarcoplasmic
reticulum from MH and control muscle. The threshold for Ca–ICaR and
the amount of Ca^{2+} released were significantly greater in sarcoplasmic
reticulum from control versus MH muscle.

In order to localize calcium in muscle, biopsies from the biceps femoris
muscle of MHS pigs were taken before and during an MH attack induced
by succinylcholine plus halothane.[29] The specimens were treated with potassium pyroantimonate. This study also measured ATP and creatine
phosphate to show that oxidative phosphorylation was not uncoupled
during the critical period. Nevertheless, during the hyperthermic period,
the muscle of MHS pigs had less ATP and creatine phosphate than did
the muscle of the control animals. The pyroantimonate method demonstrated that during the MH attack, increased amounts of calcium were
present in the mitochondria of the MHS pigs.

Endo et al.[30] studied the changes in skinned fibers of skeletal muscle of a patient with MH. In this patient's muscle, the Ca–ICaR mechanism showed a significantly higher sensitivity to Ca^{2+} than that in normal muscles, and the maximum rate of Ca^{2+} release at a sufficiently high concentration of Ca^{2+} was also significantly higher. Halothane accelerated Ca–ICaR to a similar extent in both the MH patient and normal muscles. No difference was observed in the properties of Ca^{2+} uptake by the sarcoplasmic reticulum and of the contractile protein system between the patient's and normal muscles. The authors concluded that sensitivity to the contraction–inducing action of halothane, and most probably of caffeine, must be due to labilized Ca–ICaR mechanism but not to the higher drug sensitivity in the narrow sense. In general, they suggested that in MH the Ca–ICaR mechanism was much easier to evoke than in normal subjects, and if it is further potentiated by halothane, it results in net Ca^{2+} release that causes muscle rigidity and excessive heat production.

Niebroj–Dobosz et al.[31] showed that, in animals with MH, Ca^{2+}–ATPase was not activated in the plasma membranes of the muscle fibers when compared with control animals ($n = 4$). They believed that halothane affected the cell membrane of the muscle fibers and increased its permeability. Under normal conditions, owing to the action of Ca^{2+}–ATPase in the cell membrane, the correct calcium gradient is maintained. A failure of the activity of this enzyme in MH animals would be responsible for impaired extrusion of calcium from the cell. An increase of Ca^{2+}–ATPase of the sarcoplasmic reticulum was observed in animals with MH; the opposite was noted in the control group. This increase in Ca^{2+}–ATPase would indicate an increase in the calcium transport into the sarcoplasmic reticulum. No significant changes in the activity of Mg^{2+}–ATPase of the sarcoplasmic reticulum and of Mg^{2+}–ATPase of actinomyosin were observed after exposure to halothane in control pigs and those with MH. The amount of Troponin C was significantly changed in the animals with MH. This may be connected with disturbances in the contraction cycle in porcine MH syndrome.

The Ca^{2+}–releasing action of halothane on fragmented sarcoplasmic reticulum (FSR) from bullfrog and rabbit skeletal muscle was examined in order to understand the mechanism of Ca^{2+} release in reference to the etiology of MH.[32] Halothane had a dual action on FSR: the Ca^{2+} release and the inhibition of Ca^{2+} uptake. The addition of halothane to loaded FSR caused a rapid Ca^{2+} release followed by a sluggish Ca^{2+} leakage, which was probably due to a decreased capacity for Ca^{2+} uptake. The properties of the rapid Ca^{2+} release by halothane were similar to those of caffeine. Caffeine shifted the dose–response curve for Ca^{2+} release by halothane to a steeper relation at a range of much lower concentrations and suggested that the action of halothane may not be identical with that of caffeine in spite of many similarities.

Takagi et al.[33] also demonstrated that the Ca^{2+}–releasing action of halothane on the sarcoplasmic reticulum was similar to that of caffeine.

Britt and Kalow,[34] Kalow et al.,[35] and Moulds and Denborough[36] also considered that MH was triggered by the probable Ca^{2+}–releasing action of halothane on sarcoplasmic reticulum.

Nelson[37] suggested a defect in the mechanism causing calcium release from sarcoplasmic reticulum in MH–affected muscle. He isolated two fractions of sarcoplasmic reticulum, one light (LSR) and one heavy (HSR), from gracilis muscle of control and MHS pigs. Part of the muscle biopsy was used to compare the contracture sensitivity of the muscle to the calcium–releasing effects of caffeine on isolated sarcoplasmic reticulum membrane. Gracilis muscle of MH pigs was more sensitive to the contracture–producing effects of caffeine than was control pig muscle. When LSR fractions were optimally loaded with calcium for caffeine-induced calcium release, no difference in calcium–releasing effects of varying caffeine doses was observed between MH and control LSR. The HSR fractions could not be loaded with calcium in a manner similar to the LSR fractions because of an apparent Ca–ICaR phenomenon. Therefore, calcium threshold for Ca–ICaR was compared between MH and control HSR fraction. In this study the average calcium concentration threshold for Ca–ICaR was markedly lower for MH versus control HSR. Caffeine decreased the threshold for Ca–ICaR more in the MH than in control HSR.

Allen et al.[38] measured Ca^{2+} uptake in FSR in 121 patients. The average MH sarcoplasmic reticulum Ca^{2+} uptake was 60% lower than controls, suggesting decreased uptake in MH sarcoplasmic reticulum could be a diagnostic tool.

Kim et al.[39] used several methods to trigger Ca^{2+} release: (1) addition of halothane (e.g., 0.2 mM), (2) an increase of extravesicular Ca^{2+} concentration, (3) a combination of (1) and (2), and (4) replacement of ions (potassium gluconate with choline chloride) to produce membrane depolarization. All Ca^{2+} release activities investigated in this study (1, 2, 3 and 4) had higher rates for MH sarcoplasmic reticulum than for normal sarcoplasmic reticulum. The authors concluded that the putative Ca^{2+} release channels located in the sarcoplasmic reticulum are altered in MH sarcoplasmic reticulum.

Ca^{2+} uptake and release in muscle homogenates and FSR were examined in biopsy specimens from nonsusceptible and MHS patients.[40] It was found that Ca^{2+} uptake and release were the same in both normal and MH muscle homogenates. Halothane inhibited the uptake of Ca^{2+} by the sarcoplasmic reticulum. The halothane inhibition of Ca^{2+} uptake in normal and MH sarcoplasmic reticulum was fitted to a single line with a correlation coefficient (r) of -0.958. The pH and Ca^{2+} dependence of Ca^{2+} uptake were the same for both normal and MH sarcoplasmic reticulum. The authors concluded from this study that the Ca^{2+} uptake function of the muscle from the five patients with MH who were examined was not abnormal and could not be the locus for the initiation of MH. It seemed that halothane does decrease the retention of Ca^{2+} by muscle homogenates, perhaps by increasing the permeability of the sarcoplasmic reticulum vesicles in the homogenate. Gronert et al.[41] also reported that halothane increases the permeability of porcine sarcoplasmic reticulum to Ca^{2+}.

Nelson et al.[42] showed in 1974 that MHS pig muscle contains less total calcium than does normal pig muscle. In 1975 and 1978, they suggested that the abnormal contracture and metabolic response of MHS muscle could be altered toward normal by lowering the calcium concentration of the incubation medium.[43,44] In 1981, they measured the effects of varying Ca^{2+} concentrations of the muscle bath on the total calcium content and contractility of MHS and control pig muscles. The total magnesium and calcium content of MHS muscle was lower than control muscle at each Ca^{2+} concentration tested, and it was concluded that the lower calcium content of MHS muscle may reflect a deficiency of storage sites or that a much greater Ca^{2+} extrusion mechanism exists in MHS muscle. The fact that magnesium content was also lower in MHS muscle favored a storage deficiency. It was also shown that MHS muscle had increased sensitivity to the contracture–producing effects of halothane and caffeine, but the abnormal contracture response of MHS muscle to halothane can be abolished by incubating the muscle in the absence of Ca^{2+}.[45]

Although abnormal halothane contracture could be abolished by lower Ca^{2+} concentrations, MHS muscle still retained sufficient calcium to produce marked contracture responses when exposed to caffeine. The authors concluded that a caffeine–sensitive calcium pool which remained in the muscle after incubation at low Ca^{2+} concentration may be different from the calcium pool that contributes to contracture produced by halothane alone. These data supported their previous report (1977) that halothane and caffeine have different sites of action for producing contracture. It also suggested that halothane produced contracture in MHS muscle by acting on a pool of calcium that is in equilibrium with the free Ca^{2+} concentration of the extracellular space.[45]

In their next study on the Ca^{2+} transport function of sarcoplasmic reticulum isolated from human MH skeletal muscle, Nelson et al. showed no defect in the Ca^{2+} transport function of sarcoplasmic reticulum from MHS skeletal muscle.[46]

Ohnishi[47] used sarcoplasmic reticulum prepared from MHS pigs and showed that dantrolene does not inhibit the native calcium permeability of the sarcoplasmic reticulum but inhibits only the halothane–induced increased portion of the permeability. Other inhibitors, such as tetracaine and ruthenium red, inhibit both the native permeability and the halothane–induced increased portion of the permeability. This unique nature of dantrolene may be related to its effectiveness in aborting the MH syndrome.

In another study, Niebroj–Dobósz et al.[48] examined muscle plasmalemma, which is implicated as the site responsible for the appearance of MH in human and susceptible strains of animals. In pigs with MH the activity of $(Na^+/K^+, Mg^{2+})$–ATPase, p–nitrophenylphosphatase, and Mg^{2+}–ATPase fell significantly during anesthesia. In the control group, the opposite occurred. In both of the groups tested, there was a marginal rise in the level of sialic acid. The levels of cholesterol and lyso derivatives were abnormal before the provoking agents were administered, and they changed significantly after onset of the MH syndrome. Anesthesia reduced the phospholipid level in both tested animal groups. Before and after the provoking agents, an impoverishment in the polypeptide pattern in the range between 80,000 and 30,000 daltons in MHS animals occurred. The authors postulated that in MH the macromolecular disorganization of the muscle plasma membranes indicates that defense mechanisms maintaining cell gradients do not work in the presence of provoking agents.

Next, a study was conducted to determine if the biochemical alterations of FSR produced by diethyl ether or thymol treatment were related to the membrane ultrastructure.[49] Concentrations of diethyl ether or thymol that abolished Ca^{2+} accumulating ability but did not reduce the Ca^{2+}–activated ATPase activity caused the formation of transparent patches on the surface of negatively stained vesicles of sarcoplasmic reticulum. Higher concentrations of diethyl ether or thymol caused a decrease in the Ca^{2+}–activated ATPase activity, an apparent loss of the $40A°$ subunits and an increased irregularity in the vesicle surface structure. In this study, the vesicle shapes observed in sectional preparations were also altered.

In further studies, the effect of long-chain unsaturated fatty acids (arachidonic, oleic, and linoleic acid) on calcium uptake and release by sarcoplasmic reticulum isolated from longissimus dorsi muscle was investigated using a Ca^{2+} electrode.[50] All three long-chain fatty acids stimulated the release of Ca^{2+} from sarcoplasmic reticulum when added after exogenous Ca^{2+} was accumulated by the vesicles, and also inhibited Ca^{2+} uptake when added before Ca^{2+}. This inhibitory effect on the calcium transport by fatty acids was prevented by bovine serum albumin through its ability to bind with the fatty acid. Both arachidonic and oleic acid stimulated the $(Ca^{2+} + Mg^{2+})$–ATPase activity of sarcoplasmic reticulum at low concentrations but inhibited the $(Ca^{2+} + Mg^{2+})$–ATPase activity at high concentrations.

In additional studies, the same authors showed that mitochondria isolated from longissiumus dorsi muscle of MHS pigs exhibited a calcium–stimulated phospholipase A_2 activity which liberated fatty acids from the phospholipds of mitochondrial membranes.[51] Such fatty acids induced the sarcoplasmic reticulum to release calcium, which contributed in vivo to the enhanced level of sarcoplasmic calcium observed in porcine MH. From both of these studies, they concluded that in porcine MH, unsaturated fatty acids from mitochondrial membranes released by endogenous phospholipase A_2 would induce the sarcoplasmic reticulum to release calcium.

In MH patients, in addition to a sarcolemmal/transverse tubular anomaly[52–55] that perturbs the Ca–ICaR mechanism,[56,57] an antioxidant enzyme deficiency[58-60] has been identified. Both defects result in increased intracellular Ca^{2+} [60-65] when triggered. Anesthetic agents trigger the intrinsic muscle membrane defect by an unknown mechanism[54] and the antioxidant system defect by generating free radicals.[59] Since mammalian muscle is a nonregenerating tissue, oxidative damage is cumulative. Such damage may labilize the intrinsic membrane defect.

The anesthetic-induced rise in myoplasmic Ca^{2+} induces hypermetabolism, which produces more free radicals[66] and thereby initiates a vicious cycle. The presence of an antioxidant system defect in MH individuals suggests that it has a permissive effect on development of MH.

O'Brien et al.[60] suggested that MH susceptibility results from an antioxidant system deficit unmasking or aggravating an intrinsic muscle membrane anomaly and state that a person from a family with a history of MH or unexplained anesthetic death should be considered MHS if erythrocyte osmotic fragility is abnormal and there is a mild unexplained elevation in serum creatine kinase. In this study, erythrocytes from MHS and normal dog were examined for antioxidant system deficiency; values for serum muscle enzymes, reticulocytes, and corpuscular hemoglobin were mildly elevated. Osmotic fragility was increased. A 35% glucose–6–phosphate dehydrogenase deficiency ($P < 0.001$) with a 40% compensatory increase ($P < 0.01$) in 6–phosphogluconate dehydrogenase activity was observed. The membrane Ca^{2+}–activated ATPase activity was also abnormal. Abnormal Ca^{2+} hemostasis in MH dog RBC was indicated by decreased ability to maintain low intracellular total Ca^{2+} during in vitro energy depletion and by abnormal activity and properties of membrane Ca^{2+}–ATPase.

Lopez et al.[67] used a Ca^{2+} selective microelectrode to measure in vivo the intracellular free Ca^{2+} concentration in skeletal muscle fibers of MHS and control swine. The resting membrane potential and the calcium potential were measured in superficial fibers of the tibialis anterior muscle. In the MHS animals $(Ca^{2+})i$ was 0.47 ± 0.06 μM (mean ± SEM, $n = 27$) whereas in controls it was 0.12 ± 0.01 μM ($n = 20$). There was no significant difference in the resting membrane potential between susceptible fibers and control fibers. The results supported the hypothesis that higher levels of ionized calcium in the myoplasm are associated with susceptibility to MH. In another study[68] these authors investigated the effect of dantrolene on the intracellular $(Ca^{2+})i$ in porcine MHS skeletal muscle fibers in vivo after the syndrome was triggered with halothane. When the syndrome was triggered, the $(Ca^{2+})i$ rapidly increased from 0.47 ± 0.06 μM (mean ± SEM, $n = 27$) to 10.62 ± 3.16 μM ($n = 28$). With the electrodes in place, 1 mg/Kg, dantrolene was administered intravenously. Intracellular $(Ca^{2+})i$ abruptly decreased to 0.35 ± 0.01 μM ($n = 6$) and the hypermetabolic state abated. These results showed that MH was accompanied by an increase in intracellular free $(Ca^{2+})i$ and the effect of dantrolene is associated with a decrease in ionized calcium.

The calcium ion uptake by mitochondria from skeletal muscle of susceptible animals given neuroleptic drugs and halothane was investigated and the uptake of calcium ions by mitochondria from skeletal muscle was similar for MHS pigs given neuroleptic drugs and normal pigs. The onset of MHS in the former was associated with a significant reduction in calcium ion uptake. Neuroleptics increased the binding of calcium by a preparation of mitochondria in vitro.[69] Also, premedication of MHS animals with neuroleptic drugs, such as azaperone, haloperidol, and spiperone, delays the onset of MH.[70,71]

The potential effects of the ionophore A23187, which enhances intracytoplasmic Ca^{2+}, were investigated in Pietrain pig muscle and were compared with the effects of caffeine, known to induce dose–dependent contracture in vitro in isolated muscle from human subjects with MH.[72] A23187 allowed a very clear differentiation between the muscles of normal and pathologic animals and indicated that exposure to A23187 should be added to the halothane and caffeine tests currently used to detect this disease. A23187 is a lipophilic carboxylic antibiotic that binds and transports divalent cations across both natural and artificial membrane bilayers.[73,74] For this reason, it is often used to study Ca^{2+} involvement in various cellular functions, including excitation–contraction coupling.[75] A23187 is also a useful tool for investigating the role of Ca^{2+} distrubances in certain muscle diseases.[76]

Kalow et al.[77] first demonstrated that the MH patient's muscle has a higher sensitivity than normal muscle to the contracture–inducing action of caffeine applied with or without halothane as a potentiator. Takagi et al.[78] later showed that halothane–induced contracture can be obtained in MH patients' muscle with a much lower concentration of the drug than in normal muscle. Both caffeine[78] and halothane[79] are considered to cause contracture by potentiating the Ca–ICaR mechanism in the sarcoplasmic reticulum.[80,81] Therefore, the higher sensitivity of MH patients' muscle to caffeine and halothane may suggest that the Ca–ICaR mechanism was abnormal in those muscles and that calcium release occurs too easily through this mechanism. However, an alternative possibility that the abnormality appears only in the presence of the drug, whereas the Ca–ICaR mechanism of the patient's muscle is quite normal in the absence of the drug, has not been excluded so far.[82]

Kalow et al.,[83] Okumura et al.,[84] Nelson and Flewellen,[85] and Ellis[86] believe that the direct cause of MH is an abnormal sustained increase of Ca^{2+} in the skeletal muscle fiber cytoplasm.

Caffeine acts directly on Ca^{2+} release from the sarcoplasmic reticulum and produces contracture without modifying the muscle fiber membrane potential. This is called "Endo's calcium-induced calcium release."[82] This effect has been used to detect MH in humans.[83–85]

Another study on the Ca–ICaR mechanism in MH showed a significantly higher sensitivity to calcium than that in normal muscles, and the maximum rate of calcium release at a sufficiently high concentration of calcium was also significantly higher.[87] Halothane accelerated Ca–ICaR to a similar extent in both the MH patient's and normal muscles. No difference was observed in the properties of calcium uptake by the sarcoplasmic reticulum and in the contractile protein system between the MH patient's and normal muscle. The authors assumed that sensitivity to the contracture–inducing action of halothane, and most probably of caffeine, must be due to a labilized Ca–ICaR mechanism but not to a higher drug sensitivity. The Ca–ICaR mechanism in MH patients is much easier to evoke than in normal subjects, and if it is further potentiated by halothane, it results in net calcium release, which in turn causes muscle rigidity and excessive heat production.

Discussion

After a period of 16 years of research on the role of the sarcoplasmic reticulum in MH, the data in the published reports are conflicting, and general agreement based on hard experimental evidence with quantitative measurements is lacking. The studies of Lopez et al.[67] are interesting and represent the first direct measurement of calcium ion activities in MHS swine muscle. However, there are alternate explanations for increased Ca^{2+} levels other than release from the sarcoplasmic reticulum. For instance, the calcium ions could be leaking into the muscle cells from the $10^{-3}M$ concentration in the intercellular space. Work by Williams et al.[88] demonstrates a 100– to 300–fold increase in norepinephrine levels in blood plasma, which could act as a potent hormonal driver to accelerate metabolism to a very high level. The studies of Steiss et al.[89] indicate that the myoneural junction in MHS pigs has a 1.83 times higher voltage than normal and lasts 1.12 times longer than normal. This myoneural malfunction is present in MHS pigs under resting conditions. It is conceivable that the increased myoneural junctional activity alters intracellular calcium ion activities. This notion is supported by the observation that nondepolarizing muscle relaxants such as metocurine iodide[90–92] and pancuronium[93] can be used to prevent the development of MH in MHS pigs exposed to halothane.

The Ca–ICaR mechanism may be one of the metabolic pathways altered in skeletal muscle by the MH genetic defect. However, the quantitive effect of this mechanism has not been proven sufficiently from an experimental point of view to accept it as the primary mechanism for the induction of MH and to preclude serious consideration of other equally valid postulated mechanisms as MH trigger points.

The use of dantrolene in treating MH is well established. However, the use of dantrolene per se does not define the mechanism of how MH occurs or precisely where the defective metabolic pathways are located.

Our studies with calcium channel blockers also suggest that MH can be aborted by blocking calcium channels at the plasmalemma level.[94] The calcium flux may very well be coming from the $10^{-3}M$ concentration outside the cell rather than from abnormal release from inside the muscle cell.

The apparent failure of the ^{45}Ca uptake procedure in frozen–section muscle biopsies to positively identify MH susceptible patients casts further doubt on the hypothesis that defective Ca^{2+} uptake by the sarcoplasmic reticulum is the primary metabolic defect in MH.[95]

References

1. Williams CH, Galvez TL, Brucker RF, Popinigis J, Vail WJ (1971) Malignant hyperpyrexia in swine: A genetic disease of membrane function. Fed Proc 30:1208

2. Brucker RF, Williams CH, Galvez TL (1971) The role of skeletal muscle sarcoplasmic reticulum in halothane-induced malignant hyperthermia in swine. Fed Proc 30:1208

3. Brucker RF, Williams CH, Popinigis J, Galvez TL, Vail WJ, Taylor CA (1973) In vitro studies on liver mitochondria and skeletal muscle sarcoplasmic reticulum fragments isolated from hyperpyrexic swine, in Gordon RA, Britt BA, Kalow W (eds) *Proceedings 1st International Symposium on Malignant Hyperthermia.* Springfield, IL: Charles C Thomas, pp 238–270

4. Popinigis J, Williams CH (1971) The uncoupling effect of halothane on liver mitochondria isolated from malignant hyperpyrexia swine. Fed Proc 30:1208

5. Williams CH (1972-73) Studies of heart, liver, and skeletal muscle mitochondria isolated from MHS swine (unpublished studies) 1972-1973

6. Campbell KP (1986) Protein components and their roles in sarcoplasmic reticulum function, in Entman ML, Van Winkle WB (eds) *Sarcoplasmic Reticulum in Muscle Physiology.* Boca Raton, FL: CRC Press, vol 1/3, pp 66–93

7. Meissner G (1975) Isolation and characterization of two types of sarcoplasmic reticulum vesicles. Biochem Biophys Acta 389:51

8. Ebashi S (1958) A granule-bound relaxation factor in skeletal muscle. Arch Biochem Biophys 76:410–423

9. Hasselbach W (1964) Relaxing factors and the relaxation of muscle. Prog Biophys Mol Biol 14:169–222

10. Campbell KP, MacLennan DH (1981) Purification and characterization of the 53,000 dalton glycoprotein from the sarcoplasmic reticulum. J Biol Chem 256:4626

11. MacLennan DH, Campbell KP, Reithmeier RAF (983) in Chung W (ed) *Calsequestrin in Calcium and Cell Function.* New York: Academic Press, vol.4, p 151

12. Entman ML, Van Winkle WB (1986) *Sarcoplasmic Reticulum in Muscle Function, Introduction.* Boca Raton, FL: CRC Press vol 1, pp2–13

13. Jones LR, Seiler SM, Van Winkle WB (1986) Regional differences in sarcoplasmic reticulum function, in Entman ML, Van Winkle WB (eds) *Sarcoplasmic Reticulum in Muscle Physiology.* Boca Raton, FL: CRC Press, vol 2/1, pp 5–8

14. Katz AM, Messineo F, Nash-Adler P (1986) Effects of amphiphilic substances on sarcoplasmic reticulum function, in Entman ML, Van Winkle WB (eds) *Sarcoplasmic Reticulum in Muscle Physiology*. Boca Raton, FL: CRC Press, vol 2/7, pp 128–130

15. Maher P, Singer SJ (1984) Structural changes in membranes produced by the binding of small amphipathic molecules. Biochemistry 23:232–240

16. Lain RF, Hess ML, Gertz EW, Briggs FN (1986) Calcium uptake activity of canine myocardial sarcoplasmic reticulum in the presence of anesthetic agents. Circ Res 23:597–604

17. Heffron JJA, Gronert GA (1979) Effects of halothane on calcium–binding and release by sarcoplasmic reticulum. Biochem Soc Trans 7:44–47

18. Fiehn W, Hasselbach W (1969) The effects of diethyl ether upon the function of the vesicles of sarcoplasmic reticulum. Eur J Biochem 9:574–578

19. Kalow W, Britt BA, Terreau ME, Haist C (1970) Metabolic error of muscle metabolism after recovery from malignant hyperthermia. Lancet 2:895–898

20. Ryan JF, Donlon JV, Matt RA, Bland JHL, Sreter FA, Lowenstein E (1974) Cardiopulmonary bypass in the treatment of malignant hyperthermia. N Engl J Med 290:1121–1122

21. Isaacs H, Heffron JJA (1975) Morphological and biochemical defects in muscles of human carriers of the malignant hyperthermia syndrome. Br J Anaesth 47:475–481

22. Britt BA, Kalow W, Gordon A, Humphrey JG, Reweastte NB (1973) Malignant hyperthermia: An investigation of five patients. Can Anaesth Soc J 20:431–467

23. Isaacs H, Heffron JJA, Badenhorst M (1975) Predictive tests for malignant hyperpyrexia. Br J Anaesth 47:1075–1080

24. Gronert GA, Heffron JJA, Taylor SR (1979) Skeletal muscle sarcoplasmic reticulum in porcine MH. Eur J Pharmacol 58:179–187

25. Ohnishi ST, Waring AJ, Fang SG, Horinchi K, Flick JL, Sadanga, KK, Ohnishi T (1986) Abnormal membrane properties of the sarcoplasmic reticulum of pigs susceptible to MH. Modes of action of halothane, caffeine, dantrolene, and two other drugs. Arch Biochem Biophys 246(2):294–301

26. Nelson TE, Bee DE (1979) Temperature perturbation. Studies of sarcoplasmic reticulum from MH pig muscle. J Clin Invest 64:895–901

27. Condrescu M, Lopez JR, Alamo L (1985) The high resting free calcium concentration in hyperthermic muscles might be associated to a diminished capacity of calcium uptake by the sarcoplasmic reticulum. Biophys J 47:282a

28. Nelson TE (1985) Abnormal contracture response and sarcoplasmic reticulum function in malignant hyperthermia skeletal muscle. Fed Proc 44(5):1377

29. Viering W, Stadhouders A, Ruitenbeek W, Sengers R (1984) Alteration of skeletal muscle during malignant hyperthermia. Fine structural studies and localization of calcium. Ultramicroscopy 14(4):418

30. Endo M, Yagi S, Ishizuka T, Horiuti K, Koga, Y, Amaha K (1983) Changes in the Ca-induced Ca release mechanism in the sarcoplasmic reticulum of muscle from a patient with malignant hyperthermia. Biochem Res 4(1):83–92

31. Niebroj-Dobosz I, Mayzner-Zawadzka E (1982) Experimental porcine malignant hyperthermia: The activity of certain transporting enzymes and myofibrillar calcium-binding protein content in the muscle fiber. Br J Anaesth 54:885–891

32. Ogawa Y, Kurebayashi N (1982) The Ca-releasing action of halothane on frag-
    mented sarcoplasmic reticulum. J Biochem 92:899–905
33. Takagi A, Sugita H, Toyokura Y, Endo M (1967) Malignant hyperthermia:
    Effect of halothane on single skinned muscle-fibers. Proc Japan Acad
    52(10):603–606
34. Britt BA, Kalow W (1970) Malignant hyperthermia: A statistical review. Can
    Anesth Soc J 17:316–330
35. Kalow W, Britt BA, Terreau ME, Haist C (1970) Metabolic error of muscle
    metabolism after recovery from malignant hyperthermia. Lancet 2:895–898
36. Moulds RFW, Denborough MA (1974) Biochemical basis of malignant hy-
    perpyrexia. Br Med J 2:241–244
37. Nelson TE (1983) Abnormality in calcium release from skeletal sarcoplasmic
    reticulum of pig susceptible to MH. J Clin Invest 72:862–870
38. Allen PD, Ryan JF, Jones DE, Mabuchi K, Virga A, Roberts J, Sreter F (1986)
    Sarcoplasmic reticulum calcium uptake in cryostat sections of skeletal muscle
    from malignant hyperthermia patients and controls. Muscle and Nerve,
    9(5):474–475
39. Kim DH, Sreter FA, Ohnishi ST, Ryan JF, Roberts J, Allen PD, Meszaros
    LG, Antonui B, Ikemoto N (1984) Kinetic studies of Ca^{2+} release from sar-
    coplasmic reticulum of normal and MH susceptible muscles. Biochem Biophys
    Acta 775:320–327
40. Blank TJ, Gruener R, Suffecool SL, Thompson M (1981) Calcium uptake by
    isolated sarcoplasmic reticulum examination of halothane inhibition, pH de-
    pendence and Ca^{2+} dependence of normal and MH human muscle. Anesth
    Analg 60:492–498
41. Gronert GA, Heffron JJA, Taylor SR (1979) Skeletal muscle sarcoplasmic re-
    ticulum in porcine MH. Eur J Pharmacol 58:179–187.
42. Nelson TE, Jones EW, Henrickson RL, Falk SN, Kerr DD (1974) Porcine
    malignant hyperthermia: Observations on the occurrence of pale, soft, exu-
    dative musculature among susceptible pigs. Am J Vet Res 35:347–350
43. Nelson TE, Bedell DM, Jones EW (1975) Porcine malignant hyperthermia:
    Effects of temperature and extracellular Ca concentration on halothane–induced
    contracture of susceptible skeletal muscle. Anesthesiology 42:301–306
44. Nelson TE (1978) Excitation–contraction coupling: A common etiologic path-
    way for malignant hyperthermia susceptible muscle, in Aldrete JA, Britt BA
    (eds) *Proceedings 2nd International Symposium on Malignant Hyperthermia,
    Denver, April 1977.* New York: Grune & Stratton, pp 23–26
45. Nelson TE, Chausmer AB (1981) Calcium content and contracture in isolated
    muscle of MH in pigs. J Pharmacol Exp Ther 219:107–111
46. Nelson TE, Belt MW, Kennamer DL, Winsett OE (1986) Studies on the Ca^{2+}
    transport function of sarcoplasmic reticulum isolated from human malignant
    hyperthermia skeletal muscle. Anesthesiology 64:A243
47. Ohnishi ST (1985) Dantrolene inhibits the halothane–induced increase of cal-
    cium permeability of sarcoplasmic reticulum prepared from MH pigs. Fed Proc
    44(5):1377
48. Niebroj-Dobosz I, Kwiatkowski H, Mayzner-Zawadzka E (1984) Experimental
    porcine MH: Macromolecular characterization of muscle plasma membranes.
    Med Biol 62:250–254

49. Greaser ML, Cassens RG, Hoekstra WG, Briskey EJ (1969) Effects of diethyl ethers and thymol on the ultrastructural and biochemical properties of purified sarcoplasmic reticulum fragments of skeletal muscle. Biochem Biophys Acta 193:73–81

50. Cheah AM (1981) Effects of long chain unsaturated fatty acids on the calcium transport of sarcoplasmic reticulum. Biochim Biophys Acta 648:113–119

51. Cheah KS, Cheah AM (1981) Mitochondrial Ca transport and Ca activated phospholipase in porcine MH. Biochim Biophys Acta 634:70–84

52. Gallant EM, Godt RE, Gronert GA (1979) Role of plasma membrane defect of skeletal muscle in MH. Muscle Nerve 2:491–494

53. Gruener R, Blanck TJ (1980) Volatile anesthetics and skeletal muscle: Evidence for sarcolemmal involvement in MH. Prog Anesthesiol 2:423–427

54. Gallant EM, Gronert GA, Taylor SR (1982) Cellular membrane potentials and contractile threshold in mammalian skeletal muscle susceptible to MH. Neurosci Lett 28:181–186

55. Neibroj-Dobeszi I, Mayzner-Zawadzka E (1982) Experimental porcine MH: The activity of certain transporting enzymes and myofibrillar calcium–binding protein content in the muscle fiber. Br J Anaesth 54:885–890

56. Ohnishi ST, Gronert GA (1983) The sarcoplasmic reticulum of swine with MH has abnormal calcium–induced and halothane–induced calcium release phenomena. Biophys J 41:167A

57. Nelson TE (1983) Abnormality in calcium release from skeletal sarcoplasmic reticulum of pigs susceptible to MH. J Clin Invest 72:862–870

58. Schanus EG, Schendel F, Lovrien RE, Rempel WE, McGrath C (1981) MH: Porcine erythrocyte damage from oxidative and glutathione perioxidase deficiency. Prog Clin Biol Res 55:323–336

59. Schanus EG, Lovrien RE (1982) MH in human deficiencies in the protective enzymes systems for oxidative damage. Prog Clin Biol Res 97:95–111

60. O'Brien PJ, Forsyth GW, Oleason DW, Thatte HS, Addis PB (1984) Canine MH susceptibility: Erythrocyte defects–osmotic fragility–G6PD and abnormal Ca^{2+} homeostasis. Can J Comp Med 48:381–389

61. Britt BA (1983) Malignant hyperthermia, in Orkin FK, Cooperman LH (eds) *Complications in Anesthesiology.* Philadelphia: J.B. Lippincott Co, pp 291–313

62. Harrison GG (1979) Porcine MH. Int Anesth Clin 17:25–61

63. Farber JL (1982) Membrane injury and calcium homeostasis in the pathogenesis of coagulation necrosis. Lab Invest 47:114–123

64. Shaler O, Leida MN, Hebbel RP, Jacob HS, Eaton JW (1981) Abnormal erythrocyte calcium homeostasis in oxidant–induced hemolytic disease. Blood 58:1232–1235

65. Nelson TE, Chausmer AB (1981) Calcium content and contractures in isolated muscle of MH in pigs. J Pharmacol Exp Ther 219:107–111

66. Valademirov YA, Oleneva VI, Suslova TB, Cheremisina ZP (1980) Lipid peroxidation in mitochondria membrane. Adv Lipid Res 17:173–249

67. Lopez JR, Jones D, Alamo L, Allen PD, Papp L, Gergely J, Sreter FA (1985) Free myoplasmic calcium concentration in skeletal fibers from MH susceptible swine, measured in vivo with Ca^{2+} selective micro–electrodes. Biophys J 47:313A

68. Lopez JR, Jones D, Alamo L, Allen PD, Papp L, Gergely J, Sreter FA (1985) Dantrolene reverses the syndrome of MH by reducing the level of intracellular Ca^{2+} concentration. Biophys J 47:313A

69. Somers CJ, McLaughlin JV (1982) MH in pigs: Calcium ion uptake by mitochondria from skeletal muscle of susceptible animals given neuroleptic drugs and halothane. J Comp Pathol 92:191–198

70. Ahern CP, Somers CJ, Wilson P, McLoughlin JV (1976) The prevention of acute malignant hyperthermia in halothane-sensitive Pietrain pigs by low doses of neuroleptic drugs. *Proceedings of the Third International Conference on Production Disease in Farm Animals,* Wageningen, The Netherlands: Centre for Agricultural Publishing and Documentation, pp 169–171

71. McLaughlin JV, Somers, CJ, Ahren AP, Urilson P (1978) The influence of neurolyptic drugs on the development of muscular rigidity in halothane sensitive pigs, in Lunt GG, Marchibanks RM (eds) *The Biochemistry of Myasthenia Gravis and Muscular Dystrophy.* London: Academic Press, pp 361–365

72. Reiss G, Monin G, Laner C (1986) Comparative effects of the ionophore A23187 on the mechanical response of muscle in normal Pietrain and pigs with MH. Can J Physiol Pharmacol 64:248–253

73. McLaughlin S, Eisenberg M (1975) Antibiotics and membrane biology. Ann Rev Biophys Bioeng 4:335–366

74. Malaisse WJ, Somers G, Valverde I, Couturier E (1981) Organic calcium-antagonists and calcium–ionophores. Arzneim Forsch Drug Res 31:628–633

75. Uchino M, Chou SM (1980) Effects of calcium ionophore A23187 on murine muscle. Ultrastructural and biochemical study. Proc Jpn Acad (Ser B) 56(7):480–485

76. Carpenter S, Karpati G (1979) Duchenne muscular dystrophy plasma membrane loss initiates muscle cell necrosis unless it is repaired. Brain 102:147–161

77. Kalow W, Britt BA, Terreau ME, Haist C (1970) Metabolic error of muscle metabolism after recovery from MH. Lancet 2:895–898

78. Endo M (1975) Mechanism of action of caffeine on the sarcoplasmic reticulum of skeletal muscle. Proc Jpn Acad 51:479–484

79. Takagi A, Sugita H, Toyokura Y, Endo M (1976) Malignant hyperpyrexia: Effect of halothane on single skinned muscle fiber. Proc Jpn Acad 52:603–606

80. Endo M, Tanaka M, Ogawa Y (1970) Calcium–induced release of calcium from the sarcoplasmic reticulum of skinned skeletal muscle fibers. Nature 228:34–36

81. Ford LE, Podolsky RJ (1970) Regenerative calcium release within muscle cells. Science 167:58–59

82. Endo M (1975) Mechanism of action of caffeine on the sarcoplasmic reticulum of skeletal muscle. Proc Jpn Acad 51(6):479–484

83. Kalow W, Britt BA, Richter R (1977) The caffeine test of isolated human muscle in relation to MH. Can Anaesth Soc J 24:678–694

84. Okumura F, Crocker BD, Denborough MA (1979) Identification of susceptibility to MH in swine. Br J Anaesth 51:171–176

85. Nelson TE, Flewellen EH (1979) MH: Diagnosis, treatment and investigations of a skeletal muscle lesion. Tex Rep Biol Med 38:105–120

86. Ellis FR (ed) (1981) Malignant hyperpyrexia, in *Inherited Disease and Anesthesia.* Amsterdam: Elsevier Biomedical Press, pp 163–199

87. Endo M, Yagi S, Ishizuka T, Horiuti K, Koga Y, Amaha K (1983) Changes in the calcium–induced release mechanism in the sarcoplasmic reticulum of the muscle from a patient with MH. Biomed Res 4(1):38–92

88. Williams CH, Dozier SE, Buzello W, Gehrke CW, Wong JK, Gerhardt KO (1985) Plasma levels of norepinephrine and epinephrine during malignant hyperthermia in susceptible pigs. J Chromatog 344:71–80

89. Steiss JE, Bowen JM, Williams CH (1981) Electromyographic evaluation of malignant hyperthermia susceptible pigs. Am J Vet Res 42:1175–1178

90. Williams CH, Roberts JT, Hoech GP, Waldman SD (1978) The fulminant hyperthermia-stress syndrome: Total neuromuscular blockade with dimethyl curare prevents the development of the syndrome in susceptible pigs. J Thermal Biol 3:104

91. Hoech GP, Roberts JT, Williams CH, Waldman SD, Simpson ST, Trim CM, Brazile J (1980) Prevention of porcine malignant hyperthermia with metocurine, in Cox B, Lomax P, Milton AS, Schonbaum E (eds) *Thermoregulatory Mechanisms and Their Therapeutic Implications. Proceedings 4th International Symposium on the Pharmacology of Thermoregulation, Oxford, 1979.* Basel: Karger, pp 137–141

92. Roberts JT, Williams CH, Hoech GP, Waldman SD, Brazile J, Simpson ST, Trim CM (1982) Prevention of halothane-induced porcine malignant hyperthermia by pretreatment with metocurine iodide. Anesthesiology 57(3):A224

93. Jones DE, Ryan JF, Taylor B, Papp L, Lopez JR, Alamo L, Sreter FA, Allen PD (1985) Pancuronium in large doses protects susceptible swine from halothane–induced malignant hyperthermia. Anesthesiology 63(3A):A344

94. Williams CH, Dozier SE, Ilias WK, Fulfer RT, Zukaitis MG, Hoech GP Jr (1985) Treatment of malignant hyperthermia (MH) with diltiazem. Fed Proc 44(5):1638

95. Nagarajan K, Fishbein WN, Muldoon SM, Pezeshkpour G (1987) Calcium uptake in frozen muscle biopsy sections compared with other predictors of malignant hyperthermia susceptibility. Anesthesiology 66:680–685

CHAPTER 7

In Vitro Studies of Drugs Affecting Malignant Hyperthermia Muscle

Wilfried K. Ilias, Charles H. Williams,
Susan E. Dozier, and Robert T. Fulfer

Introduction

In vitro testing of muscle specimen represents a very important tool in the diagnosis of malignant hyperthermia susceptible (MHS) people as well as in detection of MH-triggering properties of drugs and volatile anesthetics.[1] Since human muscle is not always available, muscle of inbred MH-susceptible (MHS) strains of pigs has been accepted as suitable for testing MH-triggering or suppressing properties of drugs and anesthetics. There are other in vitro animal models using frog muscle[2] or isolated rat muscle.[3] Rat hemidiaphragm sensitized by carbamylcholine has been used as an MH-mimicking in vitro model.[4] Since MH-inbred porcine muscle differs in several respects from human MH muscle,[1] it might be misleading to use muscle samples of animals that have not been proven to develop MH in vivo. The development of contracture as a result of combined exposure to caffeine and halothane may be different in rat muscle and porcine muscle, although resulting in the same effect that is, contracture.

Further problems in interpreting observations in isolated muscles stem from the fact that there are different reactions to drugs in vivo when compared to in vitro. Succinylcholine is one of the most potent trigger agents in vivo, whereas its triggering properties in vitro are questionable.[1] Similarly, carbamylcholine was proven to be a potent trigger of the MH-syndrome in isolated perfused porcine muscle,[5] but did not show potentiating effects of halothane triggered contracture when tested in vitro. In our laboratories, the same drug, when combined with halothane, was observed to induce contractures in the in vitro preparation of the rat hemidiaphragm[4] but did not show this effect when tested in porcine muscle and rat hemidiaphragms, respectively. Disparities exist in the interpretation of the actions of nondepolarizing muscle relaxants in suppressing halothane-induced contractures in vitro and also in the MH-triggering potency of catecholamines, especially norepinephrine.[1] In order to initiate further discussion, some results obtained in our laboratories are presented.

Carbamylcholine

Experiment 1

During screening of MHS pigs for experimental purposes from an MH inbred strain of Poland China pigs, 14 pigs were exposed to halothane 2% for up to 20 minutes at the age of 8 weeks. They were then marked according to their reaction and permanently identified by ear notches. Susceptibility to MH was determined by increased heart rate, increased temperature, or muscle rigor.[6] Two to six weeks later, the same pigs were anesthetized by intraperitoneal thiopental (22 mg/kg body weight). Anesthesia was maintained during ventilation with N_2O 66% in O_2 33% via orotracheal intubation (adjusted to an end expiratory CO_2 of 42 torr at STP), by increments of thiopental via ear vein. Muscle specimens were biopsied from the biceps femoris, and subsequently invasive monitoring of arterial blood pressure, central venous pressure, core temperature, and venous O_2 saturation (Opticath) was performed by cutdown to the femoral artery and femoral vein, respectively.

After a steady state of hemodynamics was recorded, the pigs were challenged with inspiratory 2% halothane until clear signs of MH (muscle rigor, increased core temperature, decreased venous O_2 saturation, increased end tidal CO_2) developed. If, after 60 minutes of halothane exposure, none of the heraldic signs of MH were visible, an intravenous injection of succinylcholine (1 mg/kg body weight) was administered as an additional trigger stimulus. Pigs who did not react to this procedure were given a second succinylcholine injection in order to confirm nonsusceptibility to MH. When the MH syndrome was clearly in progress, halothane was withdrawn, and the pigs were ventilated with pure oxygen, simultaneously cooled by surface cooling, and additionally treated with diltiazem at 2 mg/kg body weight.[7] After recovery from the MH syndrome, the sites of venous and arterial cannulation were surgically repaired, and the pigs were returned to the farm and saved for further MH experiments.

In vitro pig muscle

Immediately after excision, the muscle specimens were cut into strips of less than 2 mm in diameter and 30 to 40 mm in length, mounted into 37°C thermoregulated organ baths containing 70 ml modified (Ca 1.2 mM, Mg 1.2 mM, K 5.9 mM, P 1.2 mM, Cl 122.7 mM) Krebs–Ringer solution,[8] and aerated with 5% CO_2 in oxygen. Aeration was adjusted to maintain an organ bath pH of 7.4, with a flow rate of 500 to 600 ml/minute per organ bath. The muscle strips were stimulated with supramaximal square wave impulses (1 msec, 0.2 Hz) and pretension was adjusted until maximum twitch-amplitude had developed. Halothane 4% was added to the aeration mixture over a 20-minute period. After halothane was discontinued, the muscle strips were allowed to recover and to develop a second steady state. Carbamylcholine was added to a concentration of 50 mmol/L, and, after 20 minutes equilibration, halothane 4% was again added to the aeration for another 20-minute period. Later, 0.3 mmol/L caffeine was added to the organ bath as an additional trigger for muscle contracture.

Two MH-free animals that underwent the same procedures served as controls.

Results

Table 7.1 shows the reaction of the in vitro specimens to halothane exposure before and after addition of carbamylcholine to the organ baths. In MHS as well as in MH-free animals, carbamlycholine decreased the twitch amplitude excessively. Halothane-induced contracture was more pronounced in the absence of carbamylcholine. As expected, the two nonsusceptible muscle specimens showed no contracture at all. In Table 7.2 the correlation of the halothane inhalation test, halothane anesthesia test, breeding genetics, and halothane in vitro test are shown. In 35%, there was no correlation between the three screening methods, and in 21%, the halothane in vitro test showed a false negative result.

Experiment 2

Pig muscle

Muscle specimens were excised under thiopental anesthesia, prepared and attached to strain gauges as described. In a simultaneous run, the specimens were exposed to halothane 4% or aerated with carbogen (5% CO_2 in O_2) alone and directly stimulated in the same manner as described previously. After the muscle specimen had equilibrated, a dose-response curve for carbamylcholine was performed by adding carbamylcholine to the organ baths in increments of 10 mmol/L every 20 minutes.

Rat hemidiaphragm

Rat hemidiaphragms were excised so as to spare the phrenic nerve and immediately placed into the organ baths containing modified Krebs-Ringer solution. Indirect supramaximal stimulation via phrenic nerve was performed by square wave impulses of 0.2 msec duration and 0.2 Hz. After equilibration of the preparations, dose-response curves of carbamylcholine in the presence of 4% halothane and without halothane were performed by adding increments of 0.5 mmol/L carbamylcholine to the organ baths in intervals of 20 minutes.

TABLE 7.1. Decrease of twitch height and contracture after 50 mmol of carbamylcholine for 20 minutes followed by 4% halothane

Genetic status	●	●	●	●	●	◐	◐	◐	◐	◐	◐	◐	◐	◐	◐	○	$\bar{x} \pm SD$
Twitch Tension Decrease %	33	15	20	0	18	31	15	27	20	25	27	25	17	18	30	20	20.80 ± 8.30
Contracture Before CC	0.4	0.3	0.4	2.5	0.6	1.0	0.2	0.0	1.0	1.5	0.0	1.7	0.2	0.0	1.0	0.0	0.70 ± 0.7
Contracture After CC	0.06	0.0	0.0	0.7	0.1	0.3	0.04	0.0	0.4	0.0	0.0	0.5	0.0	0.0	0.3	0.0	0.15 ± 0.23[a]
Contracture decrease	84	100	100	70	84	66	80	0	60	100	0	60	60	0	60	0	64.57 ± 87.80

Twitch tension and contracture decrease and expressed in % of control, contracture before and after carbamylcholine (CC) expressed in grams.
[a] Significant difference, P <0.005 (t-test for paired data).
Crossbred MHS ◐
Purebred MHS ●

TABLE 7.2. Correlation of genetics, in vivo and in vitro halothane testing

Genetics	●	●	●	●	●	●	◐	◐	◐	◐	◐	◐	◐	◐	○	○	$x + SD$
Inhalation test	+	+	+	+	+	−	+	+	+	+	−	+	+	+	−	−	○
Anesthesia test	+	+	+	+	−	+	+	+	+	+	+	+	+	+	−	−	−
In vitro test	+	+	+	+	+	+	+	−	+	+	−	+	−	+	−	−	−
Contracture (g)	0.4	0.6	0.3	2.5	0.4	1.0	0.2	0.0	0.2	1.7	0.0	1.7	1.5	1.0	0.0	0.0	$0.7 + 0.7$
% Twitch tension	16	28	8	100	10[a]	7	100	8[a]	0[a]	25	0[a]	25	15	25	0	0	$24 + 33$

a No correlation.
− False negative in vitro test.
● Purebred MHS; ○ Normal (MH-free); ◐ MHS crossbred.

TABLE 7.3. Dose response of carbamylcholine in direct stimulated malignant hyperthermia-nonsusceptible pig muscle with and without 4% halothane

	n	ED[a] 10 ± SEM	ED 50 ± SEM	ED 90 ± SEM
With halothane 4%	6	13.36 ± 1.23[b]	21.40 ± 1.44[c]	35.92 ± 4.32[c]
Without halothane	6	22.01 ± 0.42	46.26 ± 2.04	97.55 ± 5.28

[a]Effective doses of carbamylcholine are in mmol/L^{-1}.
[b]Difference from control is significant, $P < 0.001$
[c]Difference from control is significant, $P < 0.0005$ (t-test, unpaired).

Results

As presented in Table 7.3, carbamylcholine decreased twitch tension of the directly stimulated pig muscle, resulting in a 100% block if the concentrations exceeded 100-mmol/L. In 12 preparations, muscle contracture as expressed by baseline shift was not observed.

The effective dose (ED) 10, ED 50, and ED 90 times were decreased significantly in those muscles that were treated with halothane when compared with the preparations aerated with halothane-free carbogen. Similar observations were made in the indirectly stimulated hemidiaphragm preparations (Figure 7.1).

The dose-response curves were significantly shifted to the left in those preparations treated with halothane. However, in the rat diaphgram muscles, none of the preparations developed contractures from exposure to carbamylcholine (Table 7.4).

The dose-response curves were analyzed by probit transformation, and statistical significance was calculated with Student's t test.

Experiment 3

Biceps muscle were removed from six MHS pigs under thiopental anesthesia, immediately cut into strips and mounted into organ baths for direct stimulation as described previously. In parallel runs, specimens taken from the same animals were treated with norepinephrine 100 µg/L or equilibrated in blank organ bath solution (control). This norepinephrine concentration was chosen because it represents a value observed in pigs during MH.[9] After 20 minutes equilibration, both preparations were exposed to halothane 4%, and after a further 20 minutes, caffeine increments of 0.3 mmol/L were added to the organ baths every 5 minutes until a final concentration of 3 mmol/L was achieved.

Results

As presented in Table 7.5, the presence of norepinephrine significantly increased the twitch tension; however, it did not induce muscle contracture.

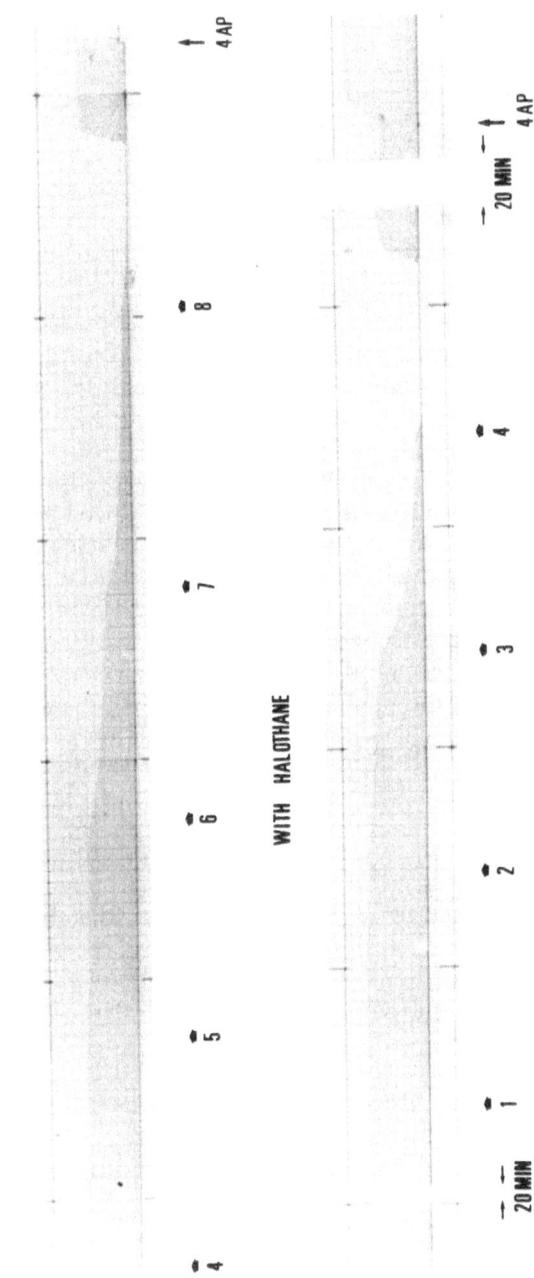

FIGURE 7.1. Original stripcharts of carbamylcholine dose-response in indirectly stimulated rat hemidiaphragms without (*upper trace*) or with halothane added to the aeration bath. Notice that the halothane preparation needed only four increments (0.5 mmol/L) to be 100% blocked (2.0 m*M*/ L final concentration), whereas the halothane free preparation was blocked after eight increments (4.0 m*M*/L final concentration). After washout, there was a hangover in twitch depression that could be antagonized by 4-aminopyridine (4 AP) at 4 mg/L.

TABLE 7.4. Dose-response of carbamylcholine in indirect stimulated rat hemidiaphragms with and without 4% halothane.

	n	ED[a] 10 ± SEM	ED 50 ± SEM	ED 90 ± SEM
With halothane 4%	6	1.099 ± 0.237[b]	1.683 ± 0.227[d]	2.677 ± 0.111[c]
Without halothane	6	1.990 ± 0.211	3.878 ± 0.460	7.842 ± 1.312

[a]Effective doses of carbamylcholine are in $mmol/L^{-1}$.
[b]Significantly different from controls, $P < 0.05$.
[c]Significantly different from controls, $P < 0.01$.
[d]Significantly different from controls, $P < 0.005$ (unpaired t-test).

During exposure to halothane alone, the norepinephrine-treated muscles reacted with significantly higher contractures than the control muscles, which, on three separate occasions, showed no contracture to halothane alone (Figure 7.2).

During the further combined halothane-caffeine stress, five of the six norepinephrine exposed muscles developed a significantly stronger contracture than the control muscles, whereas the sixth showed a reversed reaction. This might be explained by the fact that this muscle in the norepinephrine preparation responded with a contracture exceeding 30% of the twitch amplitude due to halothane alone, whereas the control muscle showed no contracture to halothane alone. This overshooting effect to halothane obviously exhausted the norepinephrine-treated muscle preparation, and it could not react to the further combined halothane-caffeine stress, whereas the control preparation showed no reaction to halothane alone and was able to respond with a very strong reaction to the combined halothane-caffeine exposure.

Experiment 4

Muscle specimens were taken from 10 MHS pigs, immediately cut into strips, and indirectly stimulated as described previously. After equilibration, 4% halothane was added to the aeration for 20 minutes. The muscle strips were then allowed to develop a second steady state; after being

TABLE 7.5. Effect of norepinephrine (NE) on halothane and halothane-caffeine induced contractures in MHS pig muscle.

	Norepinephrine n = 6		Halothane n = 6		Halothane-Caffeine n = 5	
	T.Incr. %	Contr. %	T.Incr. %	Contr. %	T.Incr. %	Contr. %
With NE	41.1 ± 2.6	0 ± 0	96.3 ± 14.3[a]	19.6 ± 4.5[a]	24.5 ± 24.5	27.2 ± 4.5[c]
Without NE	0 ± 0	0 ± 0	40.5 ± 11.2	6.9 ± 3.8	67.8 ± 30.2	19.5 ± 2.4

Mean values and SEM in increase in twitch tension (T.Incr.) and muscle tone (Contr.) expressed as % of control twitch tension.
[a]Significantly different from controls, $P < 0.05$.
[c]Significantly different from controls, $P < 0.01$ (Wilcoxon signed rank test).

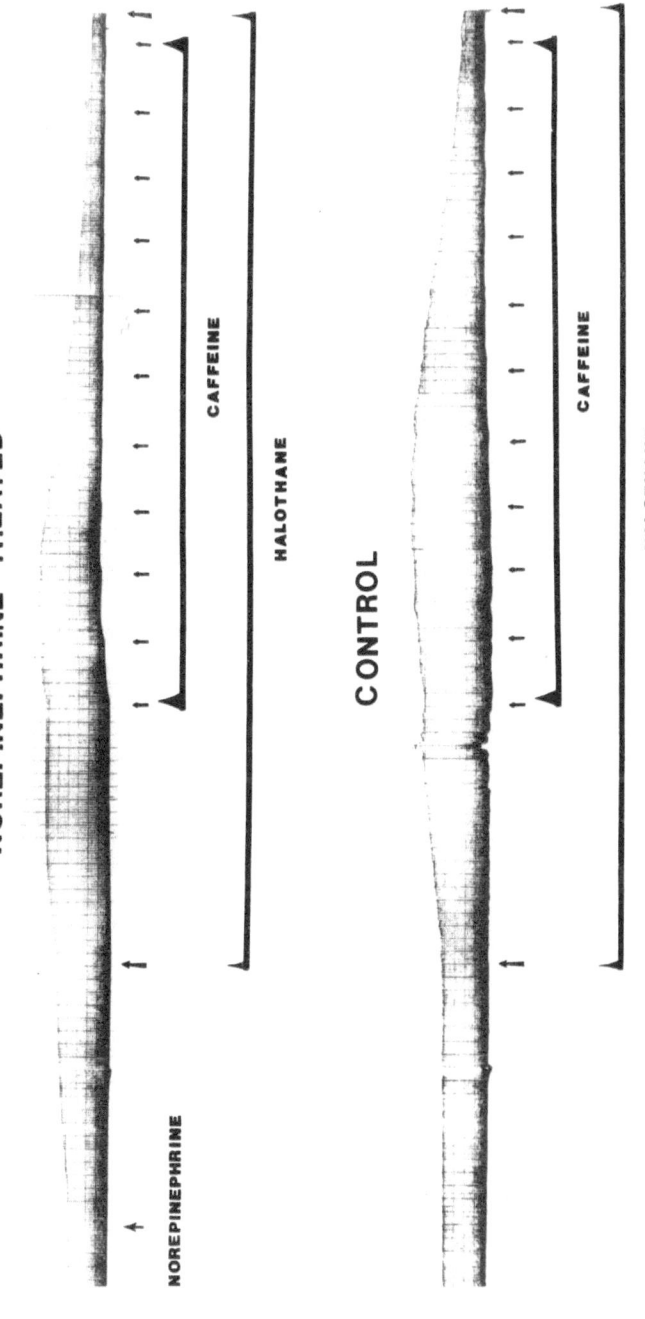

FIGURE 7.2. Original strip charts of MHS pig muscle. Norepinephrine at 100 mmol/L induced an increase in twitch tension and increased the contracture induced by caffeine increments, whereas the control preparation showed practically no effect of halothane-caffeine exposure.

equilibrated again, they were exposed to halothane for 20 minutes, and finally stressed with caffeine 0.3 mmol/L. In a parallel run, a strip from the same pig was treated with the same procedure except that diltiazem 20 μmol/L was added to the organ bath 20 minutes before the administration of halothane. Baseline shifts of more than 10% of control twitch tension, caused by halothane alone or by halothane-caffeine, were defined as being MH positive.

Results

Diltiazem decreased the control twitch tension and delayed the increase to maximum twitch tension induced by halothane as compared to the control muscles. In all preparations treated with diltiazem, a halothane contracture was suppressed, and after the halothane exposure control, twitch tension was reestablished. In contrast, the unpretreated muscles developed halothane induced contractures (Figure 7.3) and a second steady state twitch tension $26.9 \pm 5.9\%$ less than the control values. Table 7.6 shows the mean values of the maximum twitch increase. The number of baseline shifts were significantly suppressed in the diltiazem treated muscles due to halothane and combined halothane-caffeine exposure (which did not differ significantly between both groups). Time to maximum twitch increase was significantly delayed in the diltiazem treated muscles.

During screening of the MH-inbred strain of Poland China pigs, two pigs were shown repeatedly to be hyperreactive to the halothane inhalation test and the anesthesia test. In order to obtain a very sensitive muscle specimen, these two pigs were repeatedly biopsied (four times) under thiopental anesthesia. However, it was never possible to obtain viable muscle for in vitro experiments, since immediately after excision, the specimen contracted and obviously died. In attempts to suppress this contracture, the biopsied material was immediately placed in Ca^{2+}-free organ bath solution at room temperature (preoxygenated), in Mg^{2+}-enriched solutions, and even in metocurine, and procaine containing solutions. However, none of these precautions enabled us to secure a viable muscle specimen. In order to exclude preoperative stress as a possible cause of this phenomenon, these pigs were biopsied only after they had rested several days after transport from the farm to the operation theater.

Discussion

The results presented here are contrary in part to observations made by others. Carbamylcholine, presented as a possible trigger for MH-mimicking contractures in rat hemidiaphragms,[4] did not show an enhancing effect on MH triggers, such as halothane and caffeine in MHS and MH-nonsusceptible pig muscle nor in indirectly stimulated rat hemidiaphragms. The lack of contracture inducing effects in the directly stimulated pig

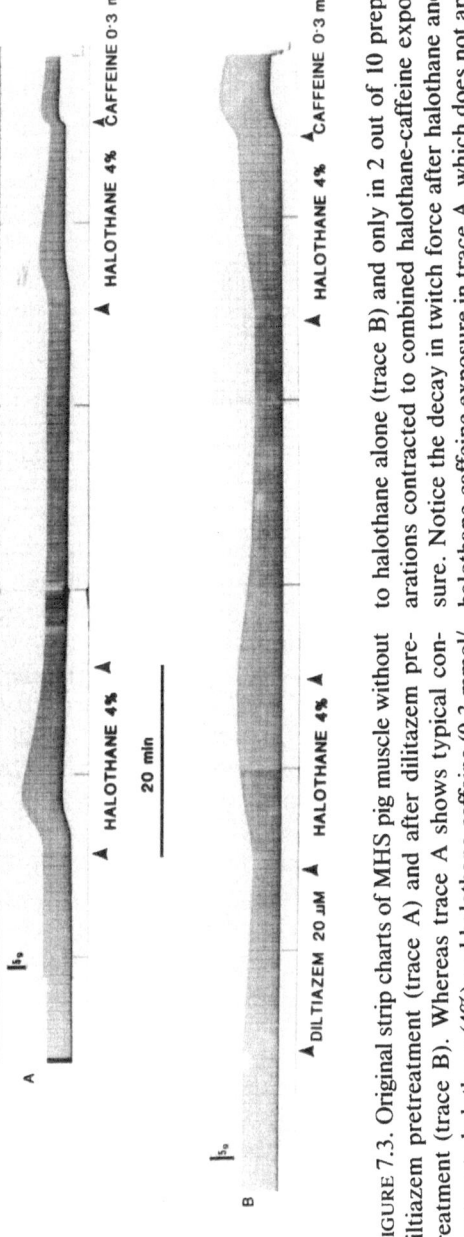

FIGURE 7.3. Original strip charts of MHS pig muscle without diltiazem pretreatment (trace A) and after diltiazem pretreatment (trace B). Whereas trace A shows typical contracture to halothane (4%) and halothane-caffeine (0.3 mmol/ L), the diltiazem treated muscle does not show contracture to halothane alone (trace B) and only in 2 out of 10 preparations contracted to combined halothane-caffeine exposure. Notice the decay in twitch force after halothane and halothane-caffeine exposure in trace A, which does not appear in the diltiazem-treated muscle (trace B).

TABLE 7.6. Effects of diltiazem pretreatment in MHS pig muscle exposed to halothane and to halothane and halothane-caffeine.

| | Twitch increase | | | |
| | Halothane 4% | | Halothane 4% + caffeine 0.3 mmol/L[b] | |
	% Control	% Diltiazem	% Control	Diltiazem
Time to maximum twitch (min)	37.9 ± 13.9	38.9 ± 06.5	19.2 ± 12.9	36.7 ± 13.0
n of baseline shifts	10	0[a]	10	2[a]

[a]Difference from control is significant, $P < 0.05$ (Wilcoxon rank and sum and Chi-square, respectively).
[b]After being equilibrated to a second steady state, muscle strips were exposed to a combined halothane-caffeine stress. (See Figure 7.3 also.)

specimen could be explained by an absence of enough end plate areas and associated sensitive membrane in the biopsied material. This explanation would account for the inconsistent contracture responses due to succinylcholine.[1] However, this explanation does not explain the lack of contractures in indirectly stimulated rat hemidiaphgrams. Although rat muscle usually does not show MH properties, the results obtained are contrary to those obtained by others.[4] As concluded from the results presented here, carbamylcholine does not seem to be a helpful adjunct in in vitro diagnosis and screening of MH. Norepinephrine, well documented to be markedly increased during MH episodes in susceptible pigs,[9] showed sensitizing effects to halothane-induced contractures in MHS pig muscle. It is well known that catecholamines are involved in the mobilization of Ca^{2+} out of membrane enclosed Ca^{2+} stores,[10] a fact that may play an important role in enhancing the action of halothane. Norepinephrine may be a helpful adjunct for the detection of MH susceptibility in vitro. Further, translating these results to the in vivo situation, there are indications that norepinephrine-induced sensitization of MH muscle due to stress exposure before halothane administration may certainly represent an important factor in inducing the MH syndrome.

Calcium entry blockers[12] may be used as a possible treatment for the MH syndrome. The results presented support this idea. Recently, the MH syndrome was treated successfully, even when fulminantly developed, in MH-inbred pigs in vivo.[7] Not only the inhibition of transmembranous Ca^{2+} transport but also the calmodulin inhibiting effects observed for some of the Ca^{2+} entry blockers may be important mechanisms involved here.[13]

In interpreting results obtained solely in in vitro observations, consideration should always be given to the fact that there are various differences among muscle reactions of isolated tissues, isolated perfused limbs, and clinical observations. The most MH-sensitive muscles from our susceptible pigs could not be tested because they died before they could be prepared for measurements. This occurred in spite of careful handling and special precautions taken with the specimen. Muscle specimens excised from MH suspect individuals, which are contracted immediately after removal and are not viable, should be handled as very suspicious for high MH susceptibility.

References

1. Gronert GA (1980) Malignant hyperthermia. Anesthesiology 53:395–423
2. Strobel GE, Bianchi CP (1971) An in vitro model of anesthetic hypertonic hyperpyrexia, halothane-caffeine induced muscle contractures: Prevention of contracture by procainamide. Anesthesiology 35:465–473
3. Harrison GG (1973) A pharmacological in vitro model of malignant hyperthermia. S Afr Med J 47:774–776
4. Fletcher JE, Rosenberg H, Hilf M (1984) Muscle contractures induced by halothane and carbamylcholine are potentiated when the agents are used in combination: Possible implications for malignant hyperthermia. Fed Proc 43:586
5. Gronert GA, Milde JH, Taylor SR (1980) Porcine muscle responses to carbachol, alpha and beta receptor agonists, halothane, or hyperthermia. J Physiol 307:319–333
6. Williams CH (1976) Some observations on the etiology of the fulminant hyperthermia-stress syndrome. Perspect Biol Med 20:120–130
7. Williams CH, Dozier SE, Ilias WK, Fulfer RT, Zukiatis MG, Hoech GP (1985) Treatment of MH (malignant hyperthermia) with diltiazem. Fed Proc 44:1638
8. Foldes FF (1981) The significance of physiological (Ca^{++}) and (Mg^{++}) for in vitro experiments on synaptic transmission. Life Sci 28:1585–1590
9. Williams CH, Dozier SE, Buzello W, Gherke CW, Wong JK, Gerhardt KO (1985) Plasma levels of norepinephrine and epinephrine during malignant hyperthermia in susceptible pigs. J Chromat: Biomed Appl 344:71–80
10. Williamson JR, Wodrow ML, Scarpa A (1975) Calcium binding to cardiac sarcolemma, in Fleckenstein A, Dhalla NS (eds) *Recent Advances in Studies on Cardiac Structure and Metabolism. Basic Function of Cations in Myocardial Activity.* Baltimore-London-Tokyo: University Park Press, vol 5, pp 61–71
11. Gallant EM, Godt RE, Gronert GA (1980) Mechanical properties of normal and malignant hyperthermia susceptible porcine muscle: Effects of halothane and other drugs. J Pharm Exp Ther 213:91–96
12. Bikhazi GB, Thomas C, Foldes FF (1979) Effects of verapamil on mammalian muscle in vitro. Anesthesiology 51:S275
13. Klumpp S, Schultz JE (1985) Calcium and calmodulin. Pharmaz Zeit 14:19–26

CHAPTER 8

The Role of the Horse in Studies Relative to Malignant Hyperthermia

Charles E. Short and Nora S. Matthews

Introduction

Malignant hyperthermia (MH) is now a well recognized syndrome in man and there are extensive efforts to understand the syndrome in both human and animal health.[1] The pig has served for a number of years as an experimental model, since a number of characteristics of MH are observed in both man and the pig, including fulminant hyperthermia, muscle rigidity, and a rapid triggering effect by either halothane anesthesia or succinylcholine.[1-4] Similar responses to circulation and ventilation have been observed in both man and swine. Malignant hyperthermia has been reported also in dogs,[5,6] horses,[7,8] and rabbits.[9]

The similarity of MH-susceptible (MHS) human subjects and susceptible horses with either exercise or anesthetic-induced myopathy is an interesting phenomenon. Waldron-Mease et al. demonstrated similar caffeine stress test responses in horses with exercise myopathy compared to human subjects with no MH susceptibility.[10] Hildebrand and Howitt observed hyperthermia in ponies.[11] Ironically, MH in the horse appears to resemble the nonhyperthermic version of MH as observed in man, with the presence of rigor and other symptoms of MH without elevated body temperature.

Hattox reported the successful use of dantrolene sodium in an MH human patient experiencing a stress-induced reaction resembling MH.[12] The patient had not had anesthesia and was taking no medications.

Walton reported that several drugs were used unsuccessfully in a patient with an unexplained muscle pain.[13] Dantrolene sodium reduced the pain slightly, but it also produced substantial muscle weakness and had to be discontinued. Relief was finally achieved with verapamil.

Dantrolene sodium has been shown to be effective in both the prevention and treatment of MH in human and porcine subjects as reported by Ryan[14] and Gronert.[15] In our laboratories, early usage in ponies demonstrated its compatability in equine anesthesia, although it was observed that higher concentrations of dantrolene sodium used in conjunction with anesthetic induction could result in hypotension and bradycardia.

Chapin et al. reported on two MH-positive pigs that died from intravenous dantrolene sodium during an MH crisis.[16] One pig received a 10 mg/kg dose as a bolus. The other received a total dose of 30 mg (0.43 mg/kg) prior to arrest. Both pigs developed severe sinus bradycardia followed by cardiac arrest from asytole during the dantrolene administration. It is suggested that the cardiovascular effects of dantrolene may be more pronounced than had been thought. There have been unscientific verbal reports of the use of dantrolene sodium for control of muscle spasms and other aspects of exercise-induced myopathies following racing, showing, or pleasure riding of horses. Therefore, we decided to evaluate various aspects of the relationship of MH to equine subjects to evaluate the safety of using dantrolene sodium in horses and to determine the responses of treatment of horses with dantrolene prior to the use of halothane anesthesia.

Materials and Methods

Three horses were located that had a previous record of extreme exercise-induced myopathy. One was a standardbred exhibiting exercise myopathy with each attempt to condition to racing form. One was a horse used for endurance riding in rough terrain that had a similar problem following prolonged exercise, and the third showed myopathy during a single physical stress event. After we determined that other indicated medications were unsuccessful, we administered 1.1 mg/kg body weight of dantrolene sodium orally per day to the three horses. The animals' performance and activities were evaluated.

The second phase of the study was to determine the compatability of prolonged use of dantrolene sodium in adult horses. Six mature female horses were included in the study. Dantrolene sodium was given SID (once a day) for 10 days at 1.1. mg/kg in three of the horses and 3.3. mg/kg in the remaining three. Blood pressures and heart rates were determined using a Dinamap (Model 1255, Critikon Inc., Tampa, FL) noninvasive blood pressure monitor before and after the 10-day period. At each recording session, three recordings for each value were determined, and means were derived for the final response.

The third phase of the study included evaluation of dantrolene sodium used in conjunction with anesthetic induction in adult horses. Fourteen horses were selected for the study. They were 1 to 16 years of age and weighed 440 to 550 kg each. Eight were females, one was an intact male, and five were male castrates. They consisted of six standardbreds, four thoroughbreds, three quarterhorses, and one appaloosa. The horses were divided into three groups. Group 1 received no dantrolene sodium. Group 2 received 500 mg of dantrolene sodium orally 2 hours prior to preanesthetic medications. Group 3 received 500 mg at 12 hours and 2 hours preanesthetic.

In addition, all horses were premedicated with acetylpromazine at 0.06 mg/kg intravenously 15 to 20 minutes before anesthetic induction with a solution consisting of 5% guaifenesin and 2 mg/ml sodium thiamylal administered intravenously until the horses were in a recumbent position and tolerated intubation.

The horses were then placed in a lateral recumbency and maintained with halothane anesthesia delivered by a North American Drager large animal anesthetic unit, using an out-of-circuit precision vaporizer and a semiclosed rebreathing system with capability of maintaining the horses on spontaneous or intermittent positive pressure ventilation. Anesthetic levels were adjusted to assure that all 14 horses were nonresponsive to the surgical stimulus. An orthopedic surgical procedure of the carpal joint was performed. Each horse received 20 ml/kg intravenous lactated Ringer's solution during the orthopedic surgical procedures. Clinical laboratory evaluations included preoperative complete blood counts, chemical profile including creatine phosphokinase (CK), and two intraoperative arterial blood gas and pH evaluations. This was followed postoperatively with a 24 hour evaluation of CK. The physiologic monitoring record was made using a multichannel recorder (model 7758B, Hewlett Packard Inc., Waltham, MA) and a Dinamap noninvasive blood pressure unit. The parameters recorded included direct and indirect arterial systolic and diastolic blood pressures, heart rates, ECG, airway flow, transthoracic pressure, and tidal volume. The transthoracic pressure was measured using an air pressure transducer connected to a thoracic cannula inserted into the pleural space. Direct blood pressure readings were made by connecting the transducer to an arterial catheter in the facial artery. All horses were maintained on foam pads during surgery or recovery periods from the time of induction of anesthesia until standing to avoid muscle damage.

Results

The results of treatment of three horses with demonstrated exercise-induced myopathy indicated that the two horses for endurance riding showed relief from the classic signs of exercise myopathy, including elevated temperature, excessive sweating, inability to move, and pain in one horse. Treatment of the horse with the pain with medications other than dantrolene sodium was ineffective in the control of the problem. Upon receiving dantrolene sodium in normal saline, the animals were more relaxed. Follow-up treatment with oral dantrolene at 1 mg/kg twice daily for 1 week followed by 1 mg/kg SID for the remainder of the treatment period was sufficient to maintain the animal without recurrence of exercise related myopathy. A similar problem in a horse used for endurance rides responded in a similar manner. The third horse which was suffering from myopathy following physical exercise, received similar treatment that was successful. The standardbred was returned to training with 1.1 mg/kg per os SID. The animal remained asymptomatic until the training program was accelerated and the animal approached the speed of 2.05 minutes on a 1 mile course.

Phase II studies to determine the tolerance of 1.1 mg/kg and 3.3 mg/kg dantrolene resulted in asymptomatic tolerance of 10 days of continuous therapy using dantrolene oral suspension. Physiologic evaluation in the horses receiving 1.1 mg/kg showed an increase in blood pressure responses at 10 days after the start of medication when compared to premedication values (Table 8.1). In contrast, at the 3.3 mg/kg level, blood pressure had dropped at 10 days when compared to control. Heart rate evaluations revealed that a slower heart rate was present after treatment at each dosage level, with a greater decrease in heart rate observed when the 3.3 mg/kg dosage level was used.

Phase III of the study did not show a significant difference in the dosage of anesthetic required for induction of anesthesia before intubation and inhalant anesthetic maintenance during orthopedic surgery. However, evaluation of cardiovascular responses revealed a lower systolic/diastolic, and mean blood pressures were observed when either 1.1 mg/kg or 2.2 mg/kg doses of dantrolene sodium was administered prior to general anesthesia. There was no significant difference in the heart rates recorded immediately after and during the maintenance of anesthesia.

The ventilatory responses were affected in these studies as shown in Table 8.2. During spontaneous respiration, the ventilatory performance of the horses had a lower than desirable pH and higher $PaCO_2$ from controls at both dosage levels of dantrolene sodium. The PaO_2 levels were adequate during spontaneous breathing in each of the three groups. However, the PaO_2 levels were significantly lower ($P < 0.05$) in controls in

TABLE 8.1. Cardiovascular responses in awake adult horses

Dosage levels	Pretreatment control values	After 10 days of dantrolene per os
		Arterial blood pressure mm Hg (torr) systolic/diastolic
1.1 mg/kg	125.3 ± 6.3	132.3 ± 24.3
	57.7 ± 10.3	59.5 ± 12.9
3.3 mg/kg	137.3 ± 15.2	128.3 ± 14.3
	57.8 ± 11.0	55.8 ± 17.8
		Mean Arterial Pressure
1.1 mg/kg	84.0 ± 12.4	88.0 ± 13.3
3.3 mg/kg	87.8 ± 7.0	83.2 ± 15.3
		Heart Rate (bpm)
1.1 mg/kg	35.2 ± 3.0	32.2 ± 1.0
3.3 mg/kg	42.3 ± 7.9	35.7 ± 4.5
		Rate Pressure Product
1.1 mg/kg	4,411	4,260
3.3 mg/kg	5,808	4,580

TABLE 8.2. Ventilation responses

	Arterial pH	$PaCO_2$	PaO_2	Bicarbonate	Total CO_2
Spontaneous breathing					
1	7.215	77	179	29	32
2	7.220	67	225	26	28
3	7.236	66	241	27	29
Mechanical respiration					
1	7.467	35	396	25	26
2	7.398	39	409	24	25
3	7.294	57	152	27	28

Group 1 controls received no dantrolene sodium.
Group 2 received 1 mg/kg dantrolene sodium as preanesthetic.
Group 3 received 2 mg/kg dantrolene sodium as preanesthetic.
Values are expressed as means.

animals receiving 1.1 mg/kg dantrolene than when maintained on mechanical ventilation. The horses in group 1 (no dantrolene) and group 2 (1.1 mg/kg dantrolene sodium) had ideal arterial blood pH, $PaCO_2$ and PaO_2 values during the surgical procedures. It was observed that arterial pH was lower, $PaCO_2$ was higher, and PaO_2 was low, although within acceptable ranges, even when mechanical ventilation was used in animals

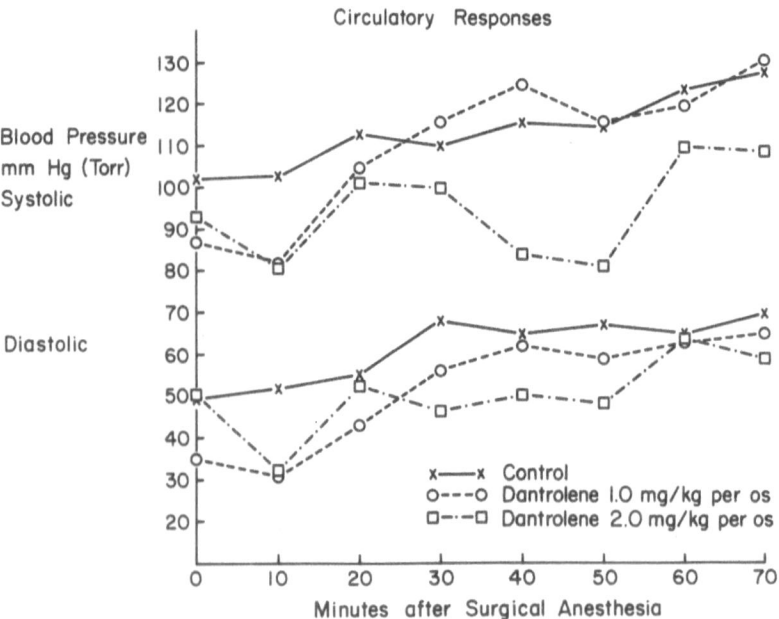

FIGURE 8.1. Illustration of systemic blood pressure during halothane anesthesia in dantrolene-pretreated horses.

receiving 2.2 mg/kg of dantrolene sodium prior to general anesthesia. The tidal volume in all horses under general anesthesia was in the range of 4.5 to 7L, airflow ranged from 0.85 to 1.5 L/second. The transthoracic pressure varied from -3 to -35 mm Hg during spontaneous respiration to +7.5 to +70 mm Hg pressure during mechanical ventilation. These values did not appear to be affected by the dosage level of dantrolene sodium. Data was analyzed using the Wilcoxon rank sum test.

All horses were maintained on general anesthesia for a minimum of 60 minutes. Blood pressures were lower in treated horses than in controls (Figures 8.1 and 8.2). All horses had an uneventful initial arousal from anesthesia during recovery. It should be noted that two horses had post-anesthetic myopathy. One in group 2 showed stiffness in the right hind leg, and one in group 3 demonstrated significant clinical myopathy during recovery and was euthanized.

The muscle enzyme activity in these horses revealed an increase in CK from the preanesthetic mean of 92 units, to a mean of 181 units 24 hours postanesthesia. In the horses receiving 1.1 mg/kg of dantrolene sodium, preanesthetic CK values were a mean of 79 units, increasing to 149 units after 24 hours postanesthesia. In the animals receiving 2.2 mg/kg of dantrolene sodium, the preanesthetic CK was 82 units, increasing to a mean of 1411 units 24 hours postanesthesia. This average was skewed by one

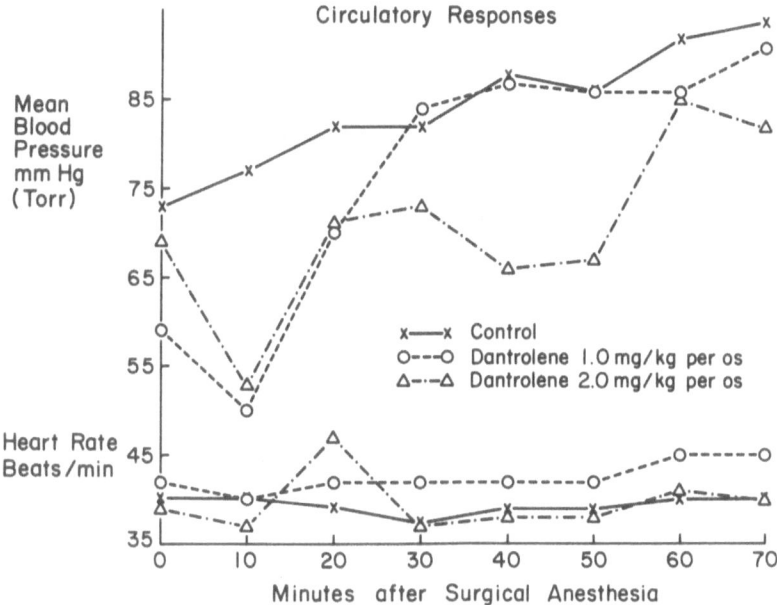

FIGURE 8.2. Illustration of cardiovascular activity during halothane anesthesia in dantrolene pre-treated horses.

animal in the group that had generalized anesthetic related myopathy and a CK value of 5325.

Discussion

Dantrolene sodium has become one of the most effective means of preventing or treating MH in pigs and humans.[17] Kolb[18] reported on 56 patients with MH who were treated with dantrolene sodium. Eighteen were treated with extemporaneous forms of dantrolene; 38 were treated with the lyophilized form of dantrolene. In the latter group, the mean dose of the drug needed to reverse the crisis was 3.35 mg/kg, with a range of 0.94 mg/kg to 15.8 mg/kg. Many were given the drug either parenterally or orally for varying periods after the MH was controlled. Thirty-two of the 38 patients recovered without sequelae. Brief summaries are presented on the 6 patients who died.

Wingard[19] discussed the use of oral dantrolene sodium to prevent MH without undesirable side effects. In normal control horses in our studies, undesirable side effects did not occur. The use of CK measurements in the horse to predict MH susceptibility is not reliable, since the elevated CK levels following exercise may give false positive results. The use of the caffeine stress test on muscle biopsies prior to anesthesia is impractical, and the platelet bioassay test[20] is still controversial among MH investigators. Since the horse seldom shows elevated body temperature as a symptom of MH, prophylactic usage of dantrolene was a possible means of control. Britt[21] suggested the use of medications that lower myoplasmic calcium to improve survival.

We studied the responses of the dosage range of 1 mg/kg to 3.3 mg/kg of dantrolene sodium in awake and anesthetized horses. Flewellen et al.[22] had reported that 2.4 mg/kg was adequate prophylaxis in humans, which was confirmed by Kerr et al.[24] Dantrolene was well tolerated in awake control horses at levels similar to those used for prophylaxis in humans.

When dantrolene was used as a preanesthetic, reduced ventilatory capability was shown during spontaneous breathing, and therefore mechanical ventilation was recommended. This confirmed the work of Oliven et al.[25] who found that there was respiratory muscle weakness as a direct response to dantrolene sodium.

Saltzman et al.[26] demonstrated that the combination of verapamil and dantrolene therapy in non-MH swine could be lethal. Our study showed that the use of dantrolene sodium in horses with known exercise-induced myopathy (probably MH) could be effective and safe but that the combination of dantrolene and halothane to produce safe anesthesia was not justified unless some means of determining MH susceptibility were used. Our horses did not demonstrate concurrent MH during anesthesia but showed evidence of postanesthetic myopathy. Grinberg et al.[27] reported latent MH in three human patients. We likewise observed delayed myopathy in horses.

Conclusion

One can conclude that anesthetic and exercise-related myopathy can occur in the horse. As a result, the horse may be an important experimental animal model for a better understanding of the entire phenomenon of anesthetic related myopathy. This study has determined that dantrolene sodium may be used successfully in some horses, unsuccessfully in others, and is contraindicated in still another group. The responses of the unanesthetized horse with exercise myopathy may improve. However, the muscle relaxing qualities of dantrolene sodium may render the horse incapable of peak performance. The use of dantrolene sodium as part of the medications being used for anesthesia induction may reduce anesthetic requirements of other drugs due to its muscle relaxing qualities. However, should hypotension occur from excessive interference with myocardial contractility, calcium chloride may be required to relieve the resulting hypotension and bradycardia. Horses tolerated both 1.1 mg/kg and 3.3 mg/kg dantrolene sodium daily for 10 days, with the exception of slight hypotension at the higher dosage levels. It should be recognized that this was not in conjunction with any other medications that may cause muscle relaxation or central nervous system depression. The physiologic response to preanesthetic administration of dantrolene sodium was partially conclusive. It was shown that it had little effect at lower dosage levels on the ventilation. Further study would be necessary to determine if higher concentrations of dantrolene sodium were responsible for the responses seen at 2.2-mg/kg. In spite of mechanical ventilation, oxygenation, and removal of carbon dioxide, the maintenance of acceptable pH levels was not accomplished in this group of animals without known susceptibility to anesthetic related myopathy. Postanesthetic myositis was not prevented in all cases.

Dantrolene sodium can be helpful in selected horses if MH susceptibility has been shown. Routine usage as a prophylactic measure to prevent postanesthetic myopathy in the horse is not indicated at this time. It is doubtful that extensive progress in this area can be accomplished until a more convenient test than the caffeine contracture method or a more reliable predicting test than evaluations of CK can be developed. The horse has demonstrated sufficient myopathy problems to make it an additional animal model to explore the mechanism MH.

Acknowledgment: Dantrolene sodium was provided for this study by Norwich Eaton Pharmaceuticals, Norwich, NY.

References

1. Short CE (1978) The significance of malignant hyperthermia in animal anesthesia, in Aldrete JA, Britt BA (eds) *Proceedings, Second International Symposium on Malignant Hyperthermia, Denver, April 1–3, 1977.* New York: Grune & Stratton, pp 175–182
2. Flewellen EH, Nelson TE (1979) Porcine malignant hyperthermia: A method to produce dantrolene prophylaxis and therapeutics. Anesthesiology 51(Suppl 3S):S248
3. Harrison G. (1977) The control and prevention of malignant hyperthermia in MHS Pigs: Some experimental observations, in Huflsz E, Sanchez-Hernandez JA, Vasconcelds G, Lunn JN (eds) *Anaesthesiology, Proceedings of the VI World Congress of Anaesthesiology, Mexico City.* Amsterdam: Excerpta Medica, pp 452–454
4. Murphy KSK, Jami L, Petit JM, Zytnicki D (1980) Differential effects of dantrolene sodium on fast and slow motor units, in *Abstracts, 54th Annual Meeting FASEB, Anaheim, CA, April 13–18.* Fed Proc 39:579
5. Dagshaw RJ, Cox RH, Rosenberg H (1981) Dantrolene treatment of malignant hyperthermia. J Am Vet Med Assoc 178:1029
6. Leary SL, Anderson LC, Manning PJ, Bache RJ, Zweber BA (1983) Recurrent malignant hyperthermia. J Am Vet Med Assoc 182:521–522
7. Short CE, White KK (1978) Anesthetic/surgical stress-induced myopathy (myositis). Part 1: Clinical occurrences, in *Proceedings 24th Annual Meeting, American Association of Equine Practitioners,* St Louis, MO: American Association of Equine Practitioners, pp 101–106
8. Manley SV, Kelly AB, Dodgson D (1983) Malignant hyperthermia-like reactions in three anesthetized horses. J Am Vet Med Assoc 183:85–89
9. Durbin CG Jr, Rosenberg H (1979) A laboratory animal model for malignant hyperpyrexia. J Pharmacol Exp Ther 210:70–74
10. Waldron-Mease E, Klein LV, Rosenberg H, Leitch M (1981) Malignant hyperthermia in a halothane anesthetized horse. J Am Vet Med Assoc 179:9
11. Hildebrand SV, Howitt GA (1983) Succinylcholine infusion associated with hyperthermia in ponies anesthetized with halothane. Am J Vet Res 44:2280–2284
12. Hattox JS (1981) Anesthesiology. J Am Vet Med Assoc 245:2182–2183
13. Walton J. Diffuse exercise-induced muscle pain of undetermined cause relieved by verapamil. Lancet 1:993
14. Ryan JF (1977) Treatment of malignant hyperthermia, in Henschel EO (ed) *Malignant Hyperthermia: Current Concepts.* New York: Appleton-Century-Crofts, Chap 3, pp 47–56
15. Gronert GA (1980) Malignant hyperthermia. Anesthesiology 53:395–423
16. Chapin JW, Chang GL, Wingard DW (1981) Asystole after intravenous dantrolene sodium in pigs. Anesthesiology 54:527–528
17. Denborough MA (1980) The pathopharmacology of malignant hyperpyrexia. Pharmacol Ther 9:357–365
18. Kolb ME (1981) Dantrolene sodium intravenous in the treatment of human malignant hyperthermia. Can J Hosp Pharm 34:47–51
19. Wingard DW (1983) Controversies regarding the prophylactic use of dantrolene for malignant hyperthermia. Anesthesiology 58:489–490

20. Jafek BW, Solomons CC, Masson NC, Mahowald MC, Gumprecht TF (1981) Current concepts of malignant hyperthermia. Otolaryngol Head Neck Surg 89:891–897

21. Britt BA (1979) Etiology and pathophysiology of malignant hyperthermia. Fed Proc 38:44–48

22. Flewellen EH, Nelson TE, Jones WP, Arens JF, Wagner DL (1983) Dantrolene dose-response in awake man: Implications for management of malignant hyperthermia. Anesthesiology 59:275–280

23. Flewellen HH, Nelson TE (1980) Dantrolene dose response in malignant hyperthermia–susceptible (MHS) swine: Method to obtain prophylaxis and therapeusis. Anesthesiology 52:303–308

24. Kerr DD, Wingard DW, Gatz EE (1977) Prevention of porcine malignant hyperthermia by oral dantrolene, in Aldrete JA, Britt BA (eds) *Proceedings, Second International Symposium on Malignant Hyperthermia, Denver, 1977.* New York: Grune & Stratton, pp 499–507

25. Oliven A, Deal EC Jr, Kelsen SG, Cherniack NS (1984) Respiratory response to partial paralysis in anesthetized dogs. J Appl Physiol 56(6):1583–1588

26. Saltzman LS, Kats RA, Corke BC, Norfleet EA, Heath KR (1984) Hyperkalemia and cardiovascular collapse after dantrolene and verapamil administration in swine, in *Abstracts, 58th Congress International Anesthesia Research Society, 1984.* Anesth Analg 63:272

27. Grinberg R, Edelist G, Gordon A (1983) Postoperative malignant hyperthermia episodes in patients who received "safe" anaesthetics. Can Anaesth Soc J 30:273–276

CHAPTER 9

Horses and Ponies as Animal Models for Malignant Hyperthermia

Susan V. Hildebrand

Introduction

The equine species, including horses and ponies, can develop myopathy associated with stress, exercise, or general anesthesia. A variety of possible causes exist, and increasing evidence indicates that malignant hyperthermia (MH) plays a role. Identification of MH susceptible (MHS) horses is important to the horse industry from both a performance and breeding standpoint, to the equine anesthesiologist, and for basic research as a new and potentially useful experimental model for further study of MH and its species variations.

At this time, investigation of equine MH is very much in the beginning stages. Most studies have focused on the biochemical and clinical aspects of equine myopathy and have not yet provided clear definition of the cause. There is a need to define the relationship between equine myopathy syndromes and MH.

Equine Myopathy—Relationship to MH

Anesthetic Related Equine Myopathy

An MH-like reaction to halothane anesthesia was first described by Klein in 1975 as a case report of a 3-year-old thoroughbred colt presented for surgical repair of a phalangeal fracture.[1] Anesthesia was uneventful until approximately 4.5 hours after induction, when the horse developed tachycardia, hypercapnia, and apparent arousal. Respiratory efforts made positive-pressure ventilation difficult. Metabolic acidosis and dark brown urine were noted. Rectal temperature rose to 108°F. Cooling efforts were unsuccessful, and the horse died.

Others have noted similar anesthetic catastrophes.[2–4] In three cases, horses received succinylcholine during halothane anesthesia. Two horses had prolonged fasciculation and failure to relax after succinylcholine.[2,3]

Rigidity was noted in one horse.[2] One horse had an uneventful recovery after treatment that included oral dantrolene.[4] Later, when recovery was complete, a muscle biopsy showed MH-like increased sensitivity in the halothane-caffeine contracture test.[4] In that horse, creatine kinase (CK) peaked at 424,750 IU 17 hours postoperatively.[4] Another horse survived anesthesia with a severe myopathy that resolved in 3 days.[3] The third horse had severe myopathy that prevented him from regaining his feet postoperatively. Therapy, including oral dantrolene, was ineffective, and the horse was euthanized the following day.[2]

Despite the clinical similarities that the above cases share with MH, such fulminant manifestations are uncommon in horses. More often, anesthetic-related myopathy is not obvious until after anesthesia, when the horse is unable to stand, has difficulty rising, or is unable to remain standing. The last may not occur until several hours after uneventful recovery. One muscle group may be affected, or severe, bilateral, generalized myopathy may be evident. Affected muscle groups appear hard, swollen, and painful, and the horse sweats profusely in obvious discomfort. Raised, hard plaques, usually on the dependent side, may appear over the ribs or masseter muscles. Myoglobinuria is frequently noted. If only one muscle group appears affected (the triceps or masseter muscle, for example), the condition usually resolves. However, more generalized muscle involvement is usually fatal. It was thought for many years that when only one limb was involved, nerve damage was to blame. Indeed, until Trimm and Mason in 1973 showed that muscle damage was responsible for this lameness, the forelimb condition was called radial paralysis.[5] Although nerve damage is possible, it appears that the most common cause of postanesthetic forelimb paresis is muscle injury. However, the cause of the muscle injury has not been easy to identify, and it has therefore been difficult to create experimentally the conditions for controlled study.

Possible Causes of Anesthetic Related Myopathy

Muscle Ischemia

The horse is a large, heavily muscled animal, and it could be predicted that anesthesia and recumbency represent a stress capable of causing injury in an otherwise normal, healthy horse. In addition, hypotension is a common problem during equine anesthesia and could easily contribute to impaired muscle perfusion. The weight of the immobile horse, combined with poor perfusion to compressed dependent muscle groups, is a logical cause of muscle damage due to ischemia. Steffey, et al. measured increases in serum enzymes associated with muscle damage in healthy horses undergoing elective anesthesia and surgical procedures of approximately 2 hours duration.[6] None of those horses suffered any clinically apparent myopathy.

Some investigators have attempted to correlate postanesthetic myopathy with intraoperative muscle ischemia, postulating a syndrome similar to the compartmental crush syndrome recognized in humans. Lindsay et al.[7] subjected six horses to deep anesthesia on a hard surface, and five of the six horses developed moderate to severe forelimb lameness, with elevations in CK that were much greater than would be expected. The lameness was not permanent or progressive, and the horses were considered normal in 48 hours. Lindsay et al. created an atypically adverse anesthetic condition, with deep anesthesia, severe hypoventilation, and a hard surface for the horses to lie on. It is believed that careful attention to positioning and padding in the form of thick mattresses, air, or waterbeds can greatly reduce the risk of muscle pressure damage during anesthesia.[8] However, Grandy et al., in an attempt to correlate arterial hypotension with postanesthetic myopathy, did a controlled study that more closely resembled equine clinical anesthesia.[9] The surface under the horse was well padded, and ventilation was supported. In the first part of the study, the horses were anesthetized with halothane, and mean blood pressure was maintained between 85 torr and 95 torr for 3 hours. Recovery was uneventful. Seven days later, the same horses were anesthetized with halothane, and mean blood pressure was maintained between 55 torr and 65 torr for 3 hours. Following the second anesthetic, all horses showed some degree of muscle injury, three of them so severe as to necessitate euthanasia. Injury was not restricted to the dependent muscle groups. The authors concluded that arterial hypotension had contributed to postanesthetic myopathy in this study. The nature of the muscle injury in the study by Grandy et al. was quite different than that found by Lindsay et al., yet Grandy's group had provided better padding for the horses. Was some other factor involved? Unfortunately, the high mortality in the Grandy et al. study meant that the most severely affected horses were lost to further investigation, and even the survivors were no longer available for study.

In an effort to better define muscle perfusion in anesthetized horses, Weaver and Lunn measured muscle perfusion using radioactive xenon.[10] Although considerable reduction in flow was noted between the awake and halothane anesthetized state, there was no difference in perfusion between the uppermost and dependent muscle groups measured. Lindsay et al. measured higher pressures in the dependent muscle compartments, but flow was not measured.[7]

There is good evidence, then, that hypotension and compression of dependent muscle groups play a role in equine anesthetic-induced myopathy. What remains unknown is the extent of that role. Can impaired muscle perfusion explain the cases in which blood pressure has been well maintained, yet the horse develops fatal generalized myopathy?

Malignant Hyperthermia and Myopathy

A second factor thought to be involved in equine postanesthetic myopathy is MH. Classic fulminant MH is rare in the anesthetized horse, but some of the myopathies associated with equine anesthesia may be an aberrant form of MH. Certainly, different manifestations of MH are well recognized in humans and swine, including nonrigid, normothermic, and delayed occurrences.[11-15] The reported incidence of equine anesthetic related myopathy from retrospective survey varies from 1% to 5% of anesthetic cases,[16.17] but it is difficult to glean from veterinary hospital records the accurate occurrence rate. A number of features associated with the incidence of equine myopathy are similar to those associated with MH.[16] Horses of the heavily muscled breeds, for example, the quarterhorses that resemble equine body builders, are thought to be more at risk, but all ranges.of myopathy, from temporary to severe fatal, have occurred in the lighter-bodied thoroughbred or standardbreds. The longer the anesthetic duration, the greater the risk, but, again, myopathies have occurred after anesthesia of less than 2 hours duration. A stormy anesthetic course with an excited induction, difficult control of blood pressure and ventilation, unexplained movement and arousal, and prolonged violent recovery warn of myopathy. Fit animals with a history of recent exertion are thought to be more at risk. However, exceptions are not uncommon. One notable case anesthetized by this author was a young light-weight thoroughbred filly undergoing a minor orthopedic procedure on her hind limb. She was in lateral recumbency. Anesthesia was of 2 hours duration and was uneventful as far as smooth induction and maintenance. Only two apparently minor problems continued throughout the procedure. Over the 2 hours, her nasopharyngeal temperature rose 1°C and her $PaCO_2$ continued to gradually rise despite increases in ventilator minute volume settings. The horse had a somewhat prolonged recovery to standing, and it was not until she stood that a severe bilateral triceps myopathy was evident. Since she was in lateral recumbency during anesthesia, pressure-perfusion mismatch to her uppermost muscle groups seems an unlikely explanation. As is often the case, this filly was lost to further study.

Exercise-related Equine Myopathy

In horses, heavy exercise beyond the anaerobic threshold is associated with muscle damage and electrolyte disturbances.[18-21] Consistent findings include elevated serum CK, hypochloremia, and depletion of muscle glycogen stores, but these do not persist in normal horses. However, exertional myopathy, known as "Tying Up Syndrome," "Azoturia," "Monday Morning Disease," or "Paralytic Myoglobinuria," can occur in some horses after mild exercise that would not normally cause muscle fatigue. Stress frequently is involved, since some horses can train at home but tie up when taken to show or other high-demand performance situations. This myopathy can vary in manifestation from gait stiffness and poor performance to severe muscle cramping, immobility, and myoglobinuria. Hypothyroidism, diet, and hormone imbalance have been implicated, but as yet the cause has not been well defined. Equine exertional myopathy has been related to MH by Rosenberg and Waldron-Mease, who compared halothane-caffeine contracture test results from muscle biopsies of five normal horses and three horses that tied up.[22] Responses from the affected horses were similar to those seen in MHS people. Waldron-Mease also reported exertional myopathy horses that returned to competitive levels of performance after therapy with dantrolene sodium.[23,24]

There are dissimilarities, too, between exertional myopathy and MH. Koterba and Carlson reported metabolic alkalosis rather than acidosis in seven horses with exertional myopathy ranging from mild to severe. All of these horses had markedly elevated serum CK.[19] Fillies may be more frequently affected than colts.[19]

Dantrolene Sodium

Dantrolene sodium is the only drug that reliably both prevents and treats MH in humans and swine. Improvement of equine anesthetic and exercise-related myopathy with dantrolene therapy suggests that there may be MH involvement in equine myopathy. Oral dantrolene therapy at 2.5 mg/kg to 4 mg/kg per os has been reported successful in treating anesthetic-related myopathy.[4,25] Low-dose oral dantrolene has also been reported successful in returning horses with exertional myopathy to normal performance.[23,24] Dantrolene was not effective given orally to one horse with severe myopathy after halothane and succinylcholine.[2] It is difficult to determine whether treatment failure in that case was due to failure of uptake of the drug from the gastrointestinal tract, whether therapy was delayed too long, whether enough drug was administered, or whether dantrolene was the appropriate choice. Other support therapy, including intravenous balanced electrolytes, analgesics, corticosteroids, and anti-inflammatory drugs, was administered to no avail.

Intravenous dantrolene treatment for equine anesthetic-related myopathy was reported by Short and White.[26] It is unclear in their report whether or not dantrolene was effective, since several other drugs were also given. Indeed, in most clinical situations dantrolene is one of many drugs given, and it is therefore difficult to determine its effect accurately. Intravenous dantrolene is preferable to oral dantrolene in an acute crisis, but at the present time, the commercial lyophilized form is cost prohibitive for routine use in the equine. The improved bioavailability of intravenous compared to oral dantrolene may be especially important to the critically ill horse.[27]

We treated one mare with a preparation of intravenous dantrolene made from dantrolene capsules.[27] In our opinion, the mare would have died with a severe postanesthetic myopathy if she had received only conventional therapy consisting of crystalloid solutions, analgesics, anti-inflammatories, and sedatives. Shortly after intravenous administration of 500 mg of dantrolene, she stopped sweating and lay quietly. Within 30 minutes, a second 500 mg was given, and 12 hours later, she was able to stand and walk normally. Unfortunately, that mare was lost to study, since she later died of surgical complications unrelated to the myopathy.

Court et al. carried out pharmacokinetic studies of both oral and intravenous dantrolene in healthy horses and ponies.[28] Their findings indicate that 2 mg/kg intravenous dantrolene was rapidly distributed from the blood to a relatively small volume of distribution, and there was rapid elimination, with a half life approximately one third that reported for humans. Uptake of 4 mg/kg dantrolene after oral administration was incomplete, with availability less than 50% of the administered dose.

More controlled studies are needed in order to assess properly the role of dantrolene in equine myopathy and for the use of dantrolene in defining the occurrence of MH in horses.

Biochemical Studies of Equine Myopathy

Cardinet et al. measured serum CK and SGOT in normal horses, both rested and after exercise, and in horses suffering from paralytic myoglobinuria.[20] Enzyme activity was also measured in specimens of normal skeletal muscle, heart muscle, brain, liver, kidney, spleen, and lung. Serum CK activity was found to be specific for skeletal and cardiac muscle necrosis in the horse. Serum CK increased with the first day of exercise but decreased over a 2-week training period. Horses with paralytic myoglobinuria had rapid and marked elevations in CK. One affected horse suffered a subclinical episode of myopathy. Although the horse was not obviously tied up, moderate elevation of CK and SGOT prompted a careful clinical examination that revealed hardness and swelling in the right gluteal. However, CK was not consistently elevated prior to exercise.

Trimm and Mason first demonstrated muscle involvement in equine postanesthetic radial paralysis by measuring increased serum CK and SGOT (now aspartate aminotransferase).[5] Those authors did a prospective study of 17 equine anesthetics and measured serum lactate and pyruvate, CK, SGOT, serum sorbital dehydrogenase (SDH), and potassium before and after anesthesia. Clinical observation of lameness was also included. Nine of the 17 horses developed lameness, 1 of them each time after three separate anesthetics. CK and SGOT were the only enzymes elevated in the lame horses. Four of the nine horses with lameness had CK and SGOT elevations consistent with those expected from general anesthesia and recumbency.[6] The other five horses had considerably higher serum enzyme values, with one horse having a value of 14,000 IU/ml CK. The latter horse was severely lame. Several of the affected horses were noted to have swollen hard plaques over the ribs or masseter muscles. Trimm and Mason suggested that limb ischemia due to impaired blood flow was the cause.

Waldron-Mease performed a similar study and did serum biochemical and urine myoglobin measurements on 12 randomly chosen horses undergoing anesthesia.[29] Samples were obtained prior to, during, and up to 72 hours after anesthesia. All animals had increased CK postoperatively compared to preoperative samples. In all but two horses, CK peaked at 6 hours postanesthesia. In those two horses, CK continued to rise for 24 hours. Myoglobin was found in the urine of five horses. One horse, whose identity in the series was not noted, had an MH-like reaction during anesthesia, with fever, resistance to mechanical ventilation, and muscle rigidity. Myoglobin was detected up to 48 hours postoperatively in that horse. Muscle stiffness and mortality data were not reported.

Johnson et al. found evidence of a 1% incidence of postsurgical myopathy in a retrospective survey of their equine patients.[17] No specific characteristics differentiated myopathy cases from unaffected horses, although there was a tendency to implicate deep, long anesthetics as contributing to myopathy. Those authors then did a prospective study, and measured preanesthetic and postanesthetic serum biochemistry. Fifteen normal horses and four horses fed a high grain diet, then subjected to hypoxia and hypercapnia during anesthesia, were studied. None of the prospectively studied horses developed myositis or any other postoperative complications. The most striking finding was a steadily increasing serum inorganic phosphate and a steady decline of total serum calcium. The authors postulated that energy depletion of muscles and ischemia due to the weight of the horse and poor perfusion were responsible. It is interesting that dietary manipulation, hypoxemia, and hypercapnia were ineffective in producing postoperative myopathy.

Muscle Biopsy Studies in Equine Malignant Hyperthermia

Halothane-Caffeine Contracture Testing of Equine Muscle

Test methods vary considerably among investigators studying humans and swine. In the veterinary literature, three reports of contracture testing of equine muscle also show variability in technique. Rosenberg and Waldron-Mease performed contracture studies on three horses with exertional myopathy and five normal horses.[22] In that study, gracilis muscle was biopsied in the three abnormal and three of the normal horses. Two of the normal horses had their quadriceps muscle biopsied. The muscle samples were clamped to maintain in vivo length. Muscle fascicles dissected from the biopsy specimens were exposed to increasing concentrations of caffeine alone, 1% halothane alone or with caffeine, or 80 mM KCl. Magnitude of contracture tension was reported. The same authors reported another contracture study, this time evaluating four horses with postanesthetic myopathy and five normal horses.[25] In that study, semimembranosus muscle was biopsied. Contracture responses in the presence of halothane at a concentration of 2% or 4%, 16 millimolar (mM) caffeine with 1% halothane, or 16 mM caffeine alone were reported. One other case report documents muscle biopsy from a horse that developed an MH-like reaction during anesthesia.[4] Muscle biopsy in that case was taken from the semimembranosus muscle. Muscle strips were exposed to halothane at 0.5% to 2% or to increasing increments of caffeine, either alone or in the presence of halothane 1%. A low threshold of contracture was found as compared to the findings of the other two studies.[22,25]

Several years ago, Hildebrand and Howitt had a serendipitous failure in a study intended to define dose-responses for succinylcholine infusion in halothane-anesthetized ponies.[30] It was not possible to achieve that goal, since four of the six ponies developed an MH-like response characterized by hyperthermia, hypertension, widening pulse pressure, and metabolic and respiratory acidosis. After initial administration of succinylcholine, the four reacting ponies had unusual, prolonged muscle fasciculation lasting up to 5 minutes. Two of those ponies developed rigidity after 1 to 2 hours of succinylcholine paralysis. The ponies all survived, and the experiments were repeated approximately 1 month later to verify the results of the first halothane-succinylcholine exposure. Results of the second challenge were similar to the first. All of the ponies survived, the only treatment in each case being cessation of anesthesia and cooling with ice packs. This was effective even after the second challenge, when the reaction was allowed to proceed until cardiovascular collapse appeared imminent. In this respect, the ponies differed from the classic outcome of MH. Serum enzymes associated with muscle damage were elevated in all of the ponies, with the reactor ponies tending to high values, but there was considerable variability.

In addition to the four reacting ponies, three horses became available for this author to study. One horse narrowly survived serious postanesthetic myopathy, one had chronic persistent tying up problems with mild exercise, and one had a reaction to halothane and succinylcholine that was similar to that seen in the ponies. We wanted to confirm MH susceptibility by halothane-caffeine contracture testing. Developing a consistent contracture test was more complicated than we anticipated from reading the literature.

We attempted to incorporate various biopsy-contracture techniques and encountered many difficulties with our early efforts.[31] Due to these

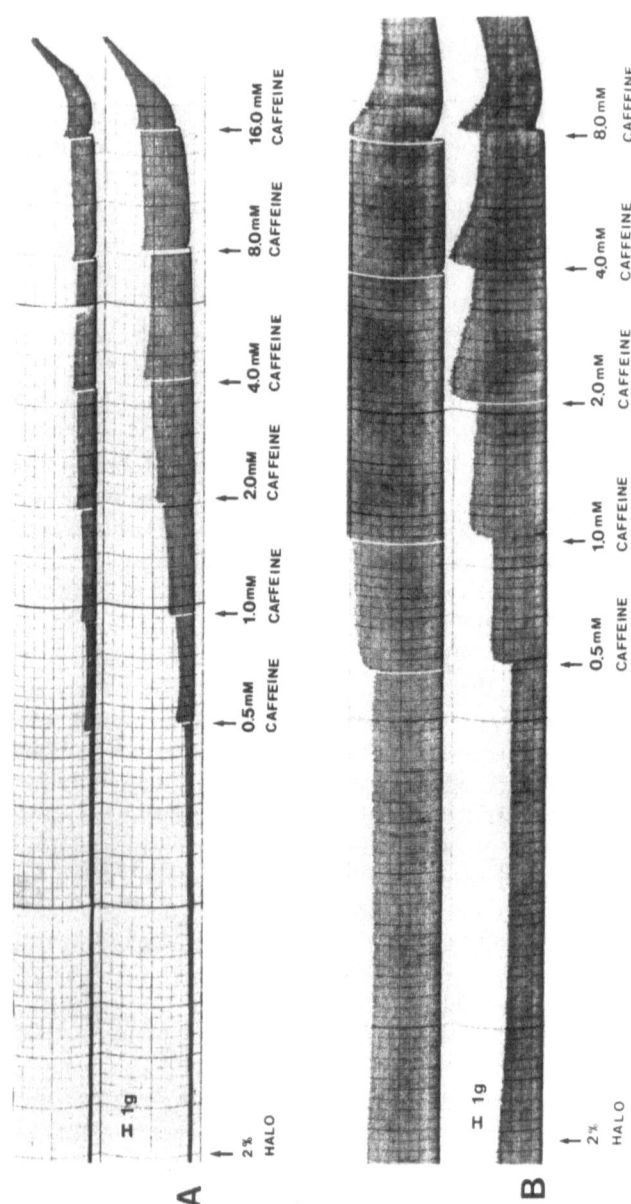

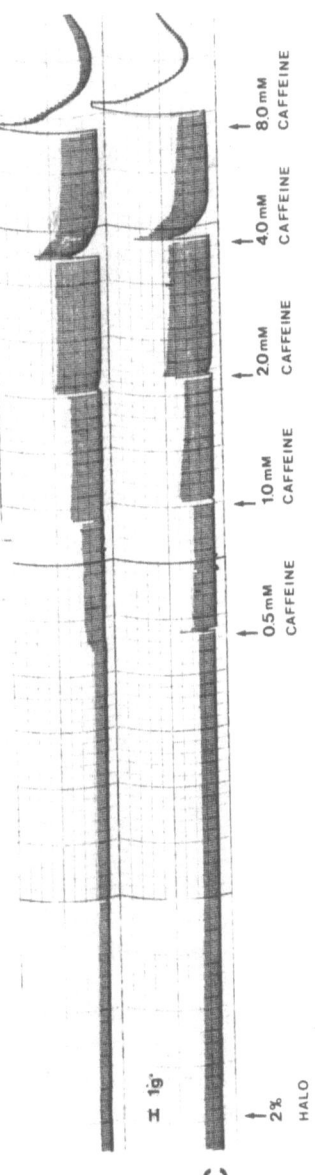

FIGURE 9.1. Traces of evoked and contracture tension responses of the equine cutaneous omobrachialis muscle to halothane and caffeine exposure in vitro. Two muscle fascicles were studied simultaneously. *A.* A typical trace showing contracture response starting at 8 m*M* caffeine plus 2% halothane. *B.* Despite differences in evoked tension response, contracture, as seen by rising baseline tension at 8 m*M* caffeine plus 2% halothane, is similar for the two muscle strips. *C.* This trace illustrates an interpertative difficulty, since early strong contracture in response to 4.0 m*M* caffeine plus 2% halothane is not sustained. Sustained contracture was seen after 8.0 m*M* caffeine in halothane. No contracture was noted in any of the traces during the 20-minute exposure to 2% halothane only.

difficulties in obtaining reliable twitch and contracture results from the semimembranosus muscle, we selected a site at the shoulder and discovered that the cutaneous omobrachialis muscle that directly overlies the lateral head of the triceps muscle was easy to work with, and a full thickness portion of the muscle belly could be taken. The surgery was accomplished with local anesthesia, using sedation with acetylpromazine alone or in combination with butorphanol. It was unclear whether other drugs commonly used for standing chemical restraint in horses, for example, xylazine or chloralhydrate with magnesium sulfate, have any effect on the subsequent response of muscle biopsy samples in contracture testing. A field block with procaine HCl 2 % provided surgical analgesia. Muscle was transported in room temperature saline to our laboratory within 10 minutes after biopsy. The muscle was held at in vivo length with special spring-loaded clamps patterned after those used by Rosenberg et al. Fascicles were dissected from the biopsy specimen, placed in a 37°C Krebs bath, and bubbled with carbogen gas. Electrical stimulus was delivered via field electrodes. In the initial protocol for exposure of the muscle to caffeine and halothane, the washout technique was used; the muscle was exposed to increasing concentrations of caffeine until the caffeine specific concentration (CSC) was reached. This was the concentration of caffeine that resulted in contracture of 1 g increase over baseline resting tension.[32] Once the CSC was achieved, the caffeine was washed out of the muscle and exposure to halothane was begun. This technique was abandoned as unsatisfactory, since the muscle returned to precaffeine resting tension after it had been exposed to the caffeine.

Fifteen normal horses have been biopsied and contracture tested as described.[31] The omobrachialis muscle was sampled and held at in vivo length. Eight fascicles, approximately $0.2 \times 0.2 \times 1$ cm, were dissected from the biopsy specimen. Two fascicles at a time were studied; they were exposed to either caffeine alone in increasing concentrations ranging from 0.5 mM to 32 mM, or to halothane 2% for 20 minutes followed by caffeine in increasing concentrations (Fig.9.1). Compared to normal CSC values for human contracture response, the omobrachialis was very resistant to the effects of caffeine and halothane with CSC ranging from 4.0 mM to 20.3 mM caffeine, with a mean of 14.03 ± 0.61 mM (SEM). In the presence of 2% halothane, CSC ranged from 4.0 mM to 8.7 mM caffeine, mean 5.6 ± 0.23 mM (SEM). In the presence of halothane alone, omobrachialis muscle from one normal horse showed contracture of 0.6 g and 0.2 g for two fascicles. To verify whether that horse would have an MH reaction to general anesthesia, he was subjected to a halothane-succinylcholine anesthetic. That challenge was uneventful, indicating that the horse was indeed normal. No other biopsies from normal horses developed contracture in the presence of halothane alone. With the established normal values for comparison, omobrachialis muscle biopsies were taken from the four reacting ponies and the three horses. The biopsy was performed twice, one taken from the left side and one from the right side. Where 14/15 normal horses showed no contracture in the presence of halothane alone, 6/7 of the reacting animals developed contracture in the presence of halothane alone. Contracture in response to halothane alone has been reported as the most sensitive indicator of MH in human muscle.[33]

Our laboratories have also examined horses with exercise-induced myopathy. The omobrachialis muscle in 14 horses was biopsied, and since exposure to caffeine alone did not seem to identify MH susceptibility in the equine test described previously, the muscle was exposed to halothane alone, either 2% or 4%, followed by increasing concentrations of caffeine. Findings from the first five horses biopsied were reported.[34] Of those five horses, one had a spontaneous contracture of 7 g, and the other had contractures of 1.0 g and 2.0 g in the presence of halothane alone. A third had an unusual contracture response in that the muscle was very resistant to the effects of halothane and caffeine, with CSC ranging from 16.0 mM to 17.3 mM caffeine. Kalow et al. have described a nonrigid form of MH in humans.[14] Muscle biopsy specimens from people so affected are resistant to the effects of halothane and caffeine in the contracture test. Possibly the horse whose muscle showed resistance would belong in such a category. It has not been possible to repeat the biopsy on that horse.

Resting CK has been elevated in all the 14 exercise myopathy horses tested. With the exception of the two positive contracture test results noted, none of the horses tested have had contracture in the presence of halothane alone. There has not been an obvious difference in CSC values between this group and the normal horses. Histopathologic and histochemical analysis revealed no abnormalities except in one horse that was finally diagnosed as having familial periodic hyperkalemic paralysis. Interestingly, in 7 of the 14 horses, the only abnormality found was elevated CK.

The breeds we have tested included Arabian and part Arabian endurance horses, thoroughbreds used for pleasure riding, and heavily muscled quarterhorses. Possibly the low number of MH positive contracture tests found by us relates to the nature of the equine population to which we have thus far had to limit our studies. Or perhaps the cutaneous omobrachialis muscle is too insensitive in its contracture responses.

In order to determine if the omobrachialis contracture response was different from that of other muscles, we have begun a comparison of the triceps muscle and the omobrachialis muscle. Histochemical analysis by Cardinet, using the method described by Cardinet and Holiday,[35] showed the omobrachialis muscle to be essentially of fiber type II makeup, whereas the triceps muscle has approximately equal representation of both types I and II. Deuster et al.[36] showed marked differences in contracture responses to caffeine and halothane between feline muscle fiber types I and II. Our goal was to determine whether such differences are relevant to equine muscle, and, if the triceps muscle was more sensitive to the effects of halothane and caffeine, whether it would be a better indicator of equine MH than the cutaneous omobrachialis muscle. Four normal horses have been biopsied, and we noted a marked difference in response between the two muscles. From the same horse, fascicles from the triceps readily develop contracture tension greater than 2 g in the presence of halothane alone, whereas the omobrachialis muscle does not contract at all. Horses whose triceps show such contracture but whose omobrachialis muscle does not have thus far reacted normally to halothane-succinyl-choline anesthetic challenge. Triceps muscle is thus far also more sensitive to the effects of caffeine in the presence of halothane than is the omobrachialis muscle. We are not certain at this point whether one muscle will be a more reliable indicator of MH susceptibility than the other.

Histochemical Analysis in Equine Malignant Hyperthermia

Histochemical studies of muscle cells, red blood cells, and platelets have not provided consistent predictive information related to MH in humans and swine.[37–45] Histochemical analysis of equine muscle to date has concentrated on defining distribution of fiber types I (slow twitch) and II (fast twitch) in normal equine muscle as it develops with age or training.[46–57] Van den Hoven et al. recently attempted to correlate muscle histology and histochemistry with tying-up syndrome.[58] They biopsied the gluteus medius muscle of 22 horses suffering from repeated attacks of exertional myopathy. Biopsies taken at onset of muscle stiffness showed increased mitochondrial ATPase activity in some fibers, and a few hours later those fibers started to lose their activity for glycogenolytic enzymes. Signs of regeneration were seen on the fourth day. By the eighth day, regenerating fibers had very active malic enzyme and decarboxylating isocitrate de-hydrogenase activities. The severity of histologic and histochemical alterations correlated well with the severity of clinical symptoms.

Andrews et al. biopsied the gluteus medius muscle and the semitendinosus muscles of one horse with exertional myopathy.[59] Histologic and histochemical changes indicated local inflammatory response. Subsarcolemmal glycogen deposits found in type II fibers suggested that impaired carbohydrate metabolism and low calcium activated myofibrillar ATPase may predispose to anaerobic metabolism. In contrast, no well-defined histochemical changes have been evident in biopsies from the cutaneous omobrachialis of horses or ponies previously described as having anesthetic-related myopathy and hyperthermia or the horses with exertional myopathy.[34] The only pathologic changes noted were in two ponies with anesthetic related hyperthermia. These were small ring myofibers with peripheral myofibrils in circular orientation (Ringbinden). In addition, one horse thought to be a normal animal proved MH positive on biopsy as determined by contracture in the presence of halothane alone in the contracture test. That horse died with severe myopathy after anesthetic challenge to verify the contracture test findings. After anesthesia, muscle biopsies from the triceps, omobrachialis, and gluteals of that horse showed massive necrosis.

Halothane-Succinylcholine Anesthetic Challenge in Equine Malignant Hyperthermia

Halothane and succinylcholine anesthetics have been administered to swine in order to cull pigs with MH susceptibility from breeding herds.[60-62] In addition, anesthetic challenge is used to verify MH-reactor swine[63,64] and dogs.[65] To my knowledge, we have conducted the only deliberate halothane-succinylcholine anesthetic challenges to verify MH-like responses in horses. We have challenged eight horses. One was the false positive described in the section on contracture testing, and four were challenged because the contracture response of the triceps muscle would have been called strongly MH positive if the muscle tested had instead been the omobrachialis. All those horses had an uneventful challenge. Two other horses had unexpected contractures of omobrachialis muscle exposed to halothane alone. One of those animals was the fatality noted in the histochemistry section. The other had prolonged fasciculations after succinylcholine and increasing hypercapnia. The challenge was terminated when the body temperature began to rise. That horse survived and will be subjected to a longer challenge at a later date. The horses with exertional myopathy are not usually available for this kind of challenge, since the owners are reluctant to subject them to a potentially life-threatening anesthetic. However, one exertional myopathy horse was donated to us, and the anesthetic challenge proceeded in similar fashion to the survival horse noted previously.

Discussion

Horses and ponies suffer from a variety of stress, exercise, and anesthetic-related myopathy. The similarities between equine myopathy and MH suggest that the equine species may prove a useful model for MH study. Muscle biopsy is relatively simple in the standing, awake horse, and ample tissue is available. However, much is unknown about equine MH. For example, hereditary aspects of the disease in the equine are as yet unknown. Hopefully, the four reactor ponies described in the halothane-caffeine contracture test section will provide future information regarding hereditary aspects of equine MH, since the four ponies comprise a stallion and three mares. The MH defect at the subcellular level has not been explored, and tissues other than muscle have not been tested. Further exploration of the equine role in MH investigation seems warranted.

References

1. Klein LV (1975) Case report: A hot horse. Vet Anesth 3:41–42
2. Manley SV, Kelly AB, Hodgson D (1983) Malignant hyperthermia-like reactions in three anesthetized horses. J Am Vet Med Assoc 183:85–89
3. Riedesel D, Hildebrand SV (1985) Unusual response following use of succinylcholine in a horse anesthetized with halothane. J Am Vet Med Assoc 187:507
4. Waldron-Mease E, Klein LV, Rosenberg H, Leitch M (1981) Malignant hyperthermia in a halothane anesthetized horse. J Am Vet Med Assoc 179:896
5. Trimm CM, Mason J (1973) Post-anaesthetic forelimb lameness in horses. Equine Vet J 5:71–76
6. Steffey EP, Farver T, Zinkl J, Wheat JD, Meagher DM, Brown MP (1980) Alterations in horse blood cell count and biochemical values after halothane anesthesia. Am J Vet Res 41:934–939
7. Lindsay W, McDonell W, Bignell W (1980) Equine postanesthetic forelimb lameness: Intracompartmental muscle pressure changes and biochemical patterns. Am J Vet Res 41:1919–1924
8. Heath RB, Redder J, Stashak T, Shaw R (1972) Protecting and positioning the equine surgical patient. Vet Med Small Anim Clin 65:1241–1245
9. Grandy JL, Steffey EP, Hodgson D, Woliner M (1987) The relationship between arterial hypotension during halothane anesthesia and the development of equine post anesthetic myopathy. Am J Vet Res 48:192–197
10. Weaver BMQ, Staddon GE (1984) Muscle perfusion in the horse. Eq Vet J 16:66–68
11. Nelson TE, Flewellen EH, Gloyna DF (1983) Spectrum of susceptibility to malignant hyperthermia—Diagnostic dilemma. Anesth Analg 62:545–552
12. Gronert GA, Thompson RL, Onofrio BM (1980) Human malignant hyperthermia awake episodes and correction by dantrolene. Anesth Analg 59:377–378
13. Beldavs J, Small V, Cooper DA, Britt BA (1971) Postoperative malignant hyperthermia: A case report. Can Anaesth Soc J 18:202–211
14. Kalow W, Britt BA, Richter, A (1977) The caffeine test of isolated human muscle in relation to malignant hyperthermia. Can Anaesth Soc J 24:678–694
15. Huckell VF, Stanloff HM, McLoughlin PR, Britt BA, Morch JE (1978) Cardiovascular manifestations of normothermic malignant hyperthermia, in Aldrete JA, Britt BA (eds) *Proceedings Second International Symposium on Malignant Hyperthermia, Denver, 1977.* New York: Grune & Stratton, pp 373–377
16. Klein L (1978) A review of 50 cases of post-operative myopathy in the horse—Intrinsic and management factors affecting risk. Am Assoc Eq Pract Proc 24:89–94
17. Johnson BD, Heath RB, Bowman B, Phillips RW, Rich LD, Voss JL (1978) Serum chemistry changes in horses during anesthesia. J Eq Med Surg 2:109–123

18. Snow DH, Kerr MG, Nimmo MA, Abbott EM (1982) Alterations in blood, sweat, urine and muscle composition during prolonged exercise in the horse. Vet Rec 110:377–384

19. Koterba A, Carlson GP (1982) Acid-base and electrolyte alterations in horses with exertional rhabdomyolysis. J Am Vet Med Assoc 180:303–306

20. Cardinet GH, Littrell JF, Freedland RA (1967) Comparative investigations of serum creatinine phosphokinase and glutamic-oxalocacetate transaminase activities in equine paralytic myoglobinuria. Res Vet Sci 8:219–226

21. Waldron-Mease E (1979) Hypothyroidism and myopathy in racing thoroughbreds and standardbreds. Eq J Med Surg 3:124–128

22. Rosenberg H, Waldron-Mease E (1977) Malignant hyperpyrexia in horses: Anesthetic sensitivity proven by muscle biopsy. Sci Abstr Am Soc Anesth, pp 333–334

23. Waldron-Mease E (1978) Correlation of postoperative and exercise-induced equine myopathy with the defect malignant hyperthermia. Am Assoc Eq Pract Proc 24:95–97

24. Waldron-Mease E (1979) Update on prophylaxis of tying-up using dantrolene. Am Assoc Eq Pract Proc 25:379

25. Waldron-Mease E, Rosenberg H (1979) Postanesthetic myositis in the horse associated with in vitro malignant hyperthermia susceptibility. Vet Sci Comm 3:45-50

26. Short CE, White KK (1977) Anesthetic/surgical stress-induced myopathy (myosites). Part I: Clinical occurrences. Am Assoc Eq Pract Proc 24:101–106

27. Gronert GA, Mansfield E, Theye RA (1978) Rapidly soluble dantrolene for intravenous use, in Aldrete JA, Britt BA (eds) *Proceedings Second International Symposium on Malignant Hyperthermia, Denver 1977*. New York: Grune & Stratton, pp 535–536

28. Court MH, Engelking LR, Dodman NH, Anwer S, Seeler DC (1987) The pharmacokinetics of dantrolene sodium in the horse. J Vet Pharmacol and Ther 10: in press.

29. Waldron-Mease E (1977) Postoperative muscle damage in horses. J Eq Med Surg 1:106–110

30. Hildebrand SV, Howitt GA (1983) Succinylcholine infusion associated with hyperthermia in ponies anesthetized with halothane. Am J Vet Res 44:2280–2284

31. Hildebrand SV, Arpin D, Howith CA (1985) Development of a technique for biopsy and in vitro contracture testing for malignant hyperthermia in equine muscle. *Proceedings, 2nd International Congress of Veterinary Anesthesia*. Santa Barbara, CA: Veterinary Practice Publishing Co., pp 168–169

32. Nelson TE, Denborough MA (1977) Studies on normal human skeletal muscle in relation to the pharmacopathology of malignant hyperpyrexia. Clin Exp Pharmacol Physiol 4:315–322

33. Ording H, Ranbler E, Fletcher R (1984) Investigation of malignant hyperthermia in Denmark and Sweden. Br J Anaesth 56:1183–1189

34. Hildebrand SV, Arpin D, Cardinet GH (1987) Exertional rhabdomyolysis related to malignant hyperthermia using the halothane-caffeine contracture test, in Gillespie JR and Robinson NE (eds) *Proceedings 2nd International Conference on Equine Exercise Physiology, Equine Exercise Physiology II*. Davis, CA: ICEEP Publications in press

35. Cardinet GH, Holliday TA (1979) Neuromuscular diseases of domestic animals: A summary of muscle biopsies from 159 cases. Ann NY Acad Sci 317:290–313
36. Deuster PA, Bockman EL, Muldoon S (1985) In vitro responses of cat skeletal muscle to halothane and caffeine. J Appl Physiol 58:521–527
37. Harriman DGF (1979) Preanesthetic investigation of malignant hyperthermia: Microscopy. Int Anesth Clin 17:97–117
38. Gronert GA (1979) Muscle contractures and adenosine triphosphate depletion in porcine malignant hyperthermia. Anesth Analg 58:367–371
39. Britt BA, Endrenyi L, Kalow W, Peters P (1976) The adenosine triphosphate (ATP) depletion test: Comparison with the caffeine contracture test as a method of diagnosing malignant hyperthermia susceptibility. Can Anaesth Soc J 23: 624–635
40. Allen PD, Ryan JF, Sreter FA, Mabuchi K (1980) Rigid vs non-rigid MH, studies of Ca^{2+} uptake and actomyosin ATPase. Anesthesiology 53:525I
41. Nagarajan K, Fishbein WN, Carlin HM, Pezeshkpour G, Muldoon SM (1985) Frozen section calcium-uptake versus halothane and caffeine contracture tests on human muscle. Anesthesiology 63(A):307
42. Blanck TJJ, Gruener R, Suffecool SL, Thompson M (1981) Calcium uptake by isolated sarcoplasmic reticulum: Examination of halothane inhibition, pH dependence and Ca^{2+} dependence of normal and malignant hyperthermia human muscle. Anesth Analg 60:492–98
43. Marjanen L, Denborough MA (1982) Adenylate kinase and malignant hyperthermia. Br J Anaesth 54:949–952
44. Giger U, Kaplan RF (1983) Halothane induced ATP depletion in platelets from patients susceptible to malignant hyperthermia and from controls. Anesthesiology 58:347–352
45. Lee MB, Adragna MG, Ewards L (1985) The use of a platelet nucleotide assay as a possible diagnosic test for malignant hyperthermia. Anesthesiology 63:311–315
46. Valbert S, Essen-Gustavsson B (1987) Metabolic response to racing determined in pools of Type I, IIA and IIB fibers and its relationship to other skeletal muscle characteristics, in Gillespie JR and Robinson NE (eds) *Proceedings 2nd International Conference on Equine Exercise Physiology, Equine Exercise Physiology II.* Davis, CA: ICEEP Publications in press
47. Valberg S, Essen-Gustavsson B, Lindholm A, Persson S (1985) Energy metabolism in relation to skeletal muscle fiber properties during treadmill exercise. Eq Vet J 17:439–444
48. Raudsepp M, Essen-Gustavsson B, Lindholm A, Persson S (1987) A field study of circulatory response and muscle characteristics in thoroughbreds during their first year of training, in Gillespie JR and Robinson NE (eds) *Proceedings 2nd International Conference on Equine Exercise Physiology, Equine Exercise Physiology II.* Davis, CA: ICEEP Publications in press
49. Bechtel P, Kline K (1987) Muscle fiber type changes in the gluteus medius of quarter and standardbred horses from birth through one year of age, in Gillespie JR and Robinson NE (eds) *Proceedings 2nd International Conference on Equine Exercise Physiology, Equine Exercise Physiology II.* Davis, CA: ICEEP Publications in press

50. Essen-Gustavsson B, Lindholm A (1985) Muscle fiber characteristics of active and inactive standardbred horses. Eq Vet J 17:434–438

51. Kline K, Laurence L, Novakovski J, Bechtel P (1987) Changes in muscle fiber type distribution within the gluteus medius of young and mature horses as a function of sampling depth, in Gillespie JR and Robinson NE (eds) *Proceedings 2nd International Conference on Equine Exercise Physiology, Equine Exercise Physiology II.* Davis, CA: ICEEP Publications in press

52. van den Hoven R, Meijer AEFH, Wensing Th, Breukink HJ (1985) Enzyme histochemical features of equine gluteus muscle fibers. Am J Vet Res 46:1755–1761

53. van den Hoven R, Wensing Th, Breukink HJ, Meijer AEFH, Kruip TAM (1985) Variation of fiber types in the triceps brachii, longissimus dorsi, gluteus medius, and biceps femoris of horses. Am J Vet Res 46:939–941

54. Snow, DH, Guy PS (1980) Muscle fiber composition of a number of limb muscles in different types of horses. Res Vet Sci 28:137–144

55. Esson B, Lindholm A, Thornton J (1980) Histochemical properties of muscle fiber types and enzyme activities in skeletal muscles of standard bred trotters of different ages. Eq Vet J 1:175–180

56. Taylor AW, Brassard L (1981) Skeletal muscle fiber distribution and area in trained and stalled standardbred horses. Southwest Vet 34:101–104

57. Barlow DA, Lloyd JM, Helhake P, Seder JA (1984) Equine muscle fiber types: A histological and histochemical analysis of selected thoroughbred yearlings. J Eq Vet Sci 2:60–66

58. van den Hoven R, Meijer AEFH, Wensing TH, Breukink HS (1987) Enzyme histochemistry in the study of equine exertional myopathy, in Gillespie JR and Robinson NE (eds) *Proceedings 2nd International Conference on Equine Exercise Physiology, Equine Exercise Physiology II.* Davis, CA: ICEEP Publications in press

59. Andrews FM, Spurgeon TL, Reed SM (1986) Histochemical changes in skeletal muscle of four male horses with neuromuscular disease. Am J Vet Res 47:2078–2083

60. Webb AJ, Gordon CHC (1978) Halothane sensitivity as a field test for stress susceptibility in the pig. Anim Prod 26:157–168

61. McGrath CJ, Lee JC, Rempel WE (1984) Halothane testing for malignant hyperthermia in swine: Dose-response effects. Am J Vet Res 45:1734–1736

62. Seeler DC, McDonell WN, Basur PK (1983) Halothane and halothane succinylcholine induced malignant hyperthermia (porcine stress syndrome) in a population of Ontario boars. Can J Comp Med 47:284–290

63. O'Brien PJ, Mickleson JR, Gronert GA, Louis CF (1985) Malignant hyperthermia susceptibility hypersensitive calcium-induced calcium release mechanism of muscle. Anesthesiology 63:270

64. Allen P, Lopez JR, Jones D, Alamo L, Papp L, Sreter FS (1985) Measurements of $(Ca^{2+})i$ in skeletal muscle of malignant hyperthermia swine. Anesthesiology 63:270

65. Cribb PH, Olfart ED, Reynolds FB (1985) The inheritance and prediction of malignant hyperthermia in the dog. *Proceedings 2nd International Congress Veterinary Anesthesia.* Santa Barbara, CA: Veterinary Practice Publishing Co., pp 170–171

CHAPTER 10

Malignant Hyperthermia in the Dog: Laboratory Investigations

Peter H. Cribb

Introduction

Malignant hyperthermia (MH) in the dog was first reported by Short and Paddleford in 1973.[1] The next reports of MH in the dog were in 1978. In that year Bagshaw et al.[2] and Cohen[3] independently reported the disease in greyhounds. Cases of MH in dogs were also reported in 1978[4] and 1981.[5] In 1981, the condition was again reported in a greyhound together with treatment with dantrolene. That the condition commonly occurs in greyhounds is well recognized in the veterinary profession.[6] A case of stress precipitated MH in a greyhound was reported in 1984.[7] Olfert et al.[8] reported the syndrome in mixed breed dogs anesthetized with halothane. The clinical syndrome described in these dogs was similar to that seen in people suffering from MH but was not confirmed by laboratory examinations.

After the death of a dog exhibiting signs of MH, the parents were traced, and from them a small breeding colony of MH susceptible (MHS) dogs was developed.[7] O'Brien et al. reported on the identification of the inherited pattern of MH susceptibility of these dogs, and their status was confirmed by halothane challenge and caffeine halothane contracture tests.[9] Inheritance of the defect conformed to a multifactorial pattern, with gradations of susceptibility. It was noted that these animals exhibited many of the physical and biochemical characteristics identified in MHS people.

These dogs were heavily muscled and were often difficult to handle, although this may have been aggravated by the fact that they were kept in a colony type situation. Two dogs died of what was assumed to be stress related causes. Subsequent testing of two of the MHS dogs at 6 years of age showed that they had an increased erythrocyte osmotic fragility in comparison to normal dogs maintained under the same conditions.[10] These dogs also showed hypertrophied muscles and had mild elevations in serum creatine kinase (CK) and aspartate transaminase. They had a mild deficiency in glucose-6-phosphate dehydrogenase and an elevation of Ca^{2+} ATPase.

Attempts to breed two MHS littermates resulted in two litters of nine and seven puppies respectively. All puppies died before 5 days of age.[11] Routine postmortem examinations were unrewarding. Similar negative findings occur in deaths in MHS neonatal pigs. A third mating resulted in pregnancy, as confirmed by ultrasonography, but fetal death occurred and the female was subsequently sterile. Subsequent mating between the MHS male and an unrelated non-MHS female produced a litter of seven live puppies. One of these died shortly after birth, and the rest were used in a study to further define the pattern of erythrocyte osmotic fragility. Biochemical testing generally showed high CK levels but there was variability in individual dogs with time. Erythrocyte osmotic fragility (EOF) testing on these dogs failed to demonstrate a specific correlation between EOF and MH susceptibility. Indeed, in this instance, changes were related to age rather than to MH susceptibility. Unfortunately, we were unable to carry out caffeine-halothane contracture tests on muscle from these dogs and diagnosis of non-MH susceptibility was based on failure of the animals to react to halothane and halothane-succinylcholine challenges.[11,12]

As in people and swine, there was a wide variation in the expression of MH in dogs, although in the laboratory animals, episodes were very severe. There was a marked increase in aerobic and anaerobic metabolism, with the production of heat, carbon dioxide, and lactate. One of the earliest signs was a marked increase in end tidal CO_2 and capnographic monitoring is very valuable in detecting the onset of an episode. Tachycardia, hyperpnea, and hyperpyrexia were severe and accompanied by cyanosis. Muscle contracture occurred early and was marked by masseter jaw spasm and elevation of the tail. Cardiac arrhythmias progress to ventricular tachycardia and arrest if the animal is untreated. End tidal CO_2 may go beyond the range of the capnograph (80 mm Hg), and arterial blood pH drops below pH 7.0. There were initial rises in serum calcium and potassium ions. Serum CK levels increased markedly. Early treatment with dantrolene, sodium bicarbonate, and ventilation with pure oxygen from a source uncontaminated by inhalation anesthetics resulted in rapid and complete recovery. The value of the routine monitoring of expired CO_2 for early detection of an impending attack cannot be overemphasized. Recovered dogs appear to suffer no ill effects if the episode is treated early.

A probable reason for the lack of popularity of the dog as a laboratory model for MH research is that the greyhound is the susceptible breed. It is a large and expensive animal to keep as a research subject. We hope to be able to crossbreed the present MH dogs to smaller breeds while at the same time retaining the genetic potential for MH. The present dogs are easy to handle and of moderate size.

References

1. Short CE, Paddleford RR (1973) Malignant hyperthermia in the dog (Letter). Anesthesiology 39:462–463
2. Bagshaw RJ, Cox RH, Knight DH, Detweiler DK (1978) Malignant hyperthermia in a greyhound. J Am Vet Med Assoc 172: 61–62
3. Cohen CA (1978) Malignant hyperthermia in a greyhound. J Am Vet Med Assoc 172:1254–1256
4. Short CE (1978) The significance of malignant hyperthermia in animal anesthesia, in Aldrete JA, Britt BA (eds) *Proceedings Second International Symposium on Malignant Hyperthermia, Denver, 1977*. New York: Grune & Stratton, 175–182
5. McGrath CJ, Rempel WE, Addis PB, Crimi AJ (1981) Acepromazine and droperidol inhibition of halothane induced malignant hyperthermia (porcine stress syndrome) in swine. Am J Vet Res 42:195–198
6. Sawyer DC (1981) Malignant hyperthermia. J Am Vet Med Assoc 179:341–344
7. Kirmayer AH, Klide AM, Purvance JE (1984) Malignant hyperthermia in a dog: Case report and review of the syndrome. J Am Vet Med Assoc 185:978–982
8. Olfert ED, White RJ, Cribb PH (1978) A malignant hyperthermia-like syndrome in the dog. Proc Can Assoc Lab Anim Sci 79:360–365
9. O'Brien PJ, Cribb PH, White RJ, Olfert ED, Steiss JE (1983) Canine malignant hyperthermia: Diagnosis of susceptibility in a breeding colony. Can Vet J 24:172–177
10. O'Brien PJ, Forsyth GW, Olexson DW, Thatte HS, Addis PB (1984) Canine malignant hyperthermia susceptibility: Erythrocyte defects, osmotic fragility, glucose-6-phosphate dehydrogenase deficiency and abnormal Ca^{2+} homeostasis. Can J Comp Med 48:381–389
11. Cribb PH, Olfert ED, Reynolds FB (1986) The inheritance and prediction of canine malignant hyperthermia susceptibility. Can Vet J 27:517–522
12. Cribb PH, Olfert ED, Reynolds FB (1985) The inheritance and prediction of malignant hyperthermia in the dog, in *Proceedings 2nd International Congress of Veterinary Anesthesia, Sacramento, CA*, Santa Barbara, CA: Veterinary Practice Publishing Co., 170–171

CHAPTER 11

Laboratory Methods for Malignant Hyperthermia Diagnosis

Jeffrey E. Fletcher and Henry Rosenberg

Introduction

A Need for Diagnostic Tests

The potentially fatal and relatively unpredictable nature of malignant hyperthermia (MH) has made it desirable to have patients exhibiting signs suggestive of MH during anesthesia or with a family history of MH screened for MH susceptibility. A number of diagnostic tests have been used for the determination of MH susceptibility. The majority of these tests use skeletal muscle, since the origin of MH may be peripheral.[1–6] To obviate the need for a muscle biopsy, less invasive tests are under development that presume that the biochemical disorder in muscle may not be organ specific and, therefore, exists in other cell types. To date, none of the less invasive tests have been widely accepted or validated by the halothane and caffeine contracture tests.[3,7]

Brief History of Test Development

The contracture response (increase in resting tension) of skeletal muscle upon exposure to halothane or caffeine is the most widely accepted method of MH diagnosis. Muscle contracture testing was pioneered by the Kalow and Britt group.[8] They reported that skeletal muscle from MH-susceptible (MHS) patients had a lower contracture threshold to caffeine than did that from controls. In addition, they found a potentiating action of halothane on the caffeine dose–response curve that also appeared to differentiate MHS from normal patients. Ellis et al.[9] demonstrated a contracture response to halothane alone in striated muscle from an MH patient. This response was not observed in control muscle. Ellis and Harriman[10] reported a second dynamic type of halothane contracture test that appeared specific for the nonrigid type of MH. Although one group has reported a contracture response to succinylcholine unique to MHS patients,[11] these succinylcholine-induced contractures have not been verified in subsequent studies.[12–15] Halsall and Ellis[14] have reported contractures synergistically induced by the combination of succinylcholine and caffeine that were antagonized by dantrolene. We recently reported on a similar synergistic interaction between halothane and succinylcholine[12,16,17] that was also antagonized by dantrolene (1 μM) in preparations from rats[16] and humans.[12] Contractures of MH muscle to KCl have also been reported.[11] With the number of different contracture tests in use by various groups, some controversy has arisen about the validity of some or all of them.

Two other diagnostic tests have been in use for several years but in only one or two laboratories, since other laboratories have not been able to obtain satisfactory results when compared to contracture testing. The Ca^{2+} uptake test[18] requires muscle biopsies but requires relatively little tissue. The test allows samples to be sent from the hospital where the biopsy is obtained to a testing laboratory. The platelet nucleotide depletion test[19] which requires a blood sample is a less invasive means of diagnosis.

Some older tests proposed for diagnosis of MH and no longer accepted as valid are not covered here, as they are reviewed adequately elsewhere.[3] Other older tests, although not believed to be valid, are presented to give a better perspective on what tissues and mechanisms have been examined.

Toward Establishing Testing Standards

An extensive list of references for contracture tests has been provided by Gronert,[4] who addresses the interlaboratory differences in contracture test results. Although there is controversy concerning its validity, the most widely employed diagnostic screen for MH was the muscle contracture test,[7] and even this test has several variants. Basically, there are two requirements for diagnostic tests used for MH. First, the test should be sensitive enough to detect all patients susceptible to MH. Second, it must be specific; that is, normal patients should not exhibit a positive response. The static halothane contracture test is generally regarded as a sensitive and specific means of diagnosis for MH.[20,21] The dynamic test was believed by some to have even greater sensitivity than the static test.[10,22] The caffeine contracture test was less sensitive than the static halothane contracture test but appeared to be very specific for MH.[20,21,23] Much less specific for MH was the halothane–caffeine specific concentration (HCSC) test,[21,24] which appeared to detect a population (phenotype-K)[15,25] whose susceptibility to MH was unclear. Because a large number of normal patients (3% to 10%)[2] to 25%[21,24] were judged positive by this test, its diagnostic value was questionable.

The static halothane contracture test has a good basis for validity. Moulds and Denborough[11] observed contractures to halothane in four survivors of MH episodes. Rosenberg and Reed[21] reported that preparations from 15 normal patients had contractures <0.4 g to halothane 2%, whereas, preparations from 6 patients with documented episodes of MH all exhibited contractures of 0.5 g or greater. Fiber bundles from an even larger normal population (24 patients) did not exhibit contractures at halothane concentrations of 2% or lower.[20] In contrast to the normal tissue, preparations from all 9 patients with documented fulminant MH reactions exhibited contractures to halothane at 2%. The same group[22] later reported on an additional 12 normal patients whose skeletal muscle did not exhibit contractures to halothane. Britt et al.[26] reported no significant contractures for preparations from 20 normal patients and contractures to halothane in those from the 13 patients with the most severe signs of MH during previous anesthesia. A study of normal rectus abdominus muscle from 57 patients demonstrated contractures greater than or equal to 0.5 g to an unspecified concentration of halothane (between 0.4% and 4.0%) in 3% of the fiber bundles tested.[27] Others[20] have reported relatively large (0.7 g) contractures in normal muscle, but this was rare and only in response to high (4%) concentrations of halothane. More controlled studies can be conducted in the porcine model of MH. Nelson et al.[25,28] demonstrated that normal porcine muscle does not exhibit contractures when exposed to halothane. In agreement with the studies in humans, muscle from MHS swine does exhibit contractures to halothane. Therefore, the contracture response to halothane does not occur in normal muscle from either humans or pigs to any great frequency, in agreement with the predicted incidence of MH in humans of between 1:7000 and 1:150,000[29] and unlike the frequently observed low HCSC in normal human muscle.

The first attempt to test a standard protocol for MH diagnosis[21] reported that the static halothane contracture test and the response to caffeine 2 mM should be used to diagnose MH, with a positive response to either signifying susceptibility. The caffeine test was added since it has considerable diagnostic value. The European Malignant Hyperthermia Group has proposed that positive results must be obtained with both the halothane contracture test and the caffeine contracture test for a positive diagnosis of MH susceptibility.[30,31] Those people who were positive in only one test were designated equivocal.

Tests Using Muscle

Contracture Tests: Considerations

Spectrum of Response

We have observed a continuum of response to halothane (Figure 11.1) similar to that reported by others.[25,26,28] Such a spectrum makes it impossible to know exactly where to draw the line distinguishing MH-negative (MH−) from MH-positive (MH+) patients. That is, two groups of

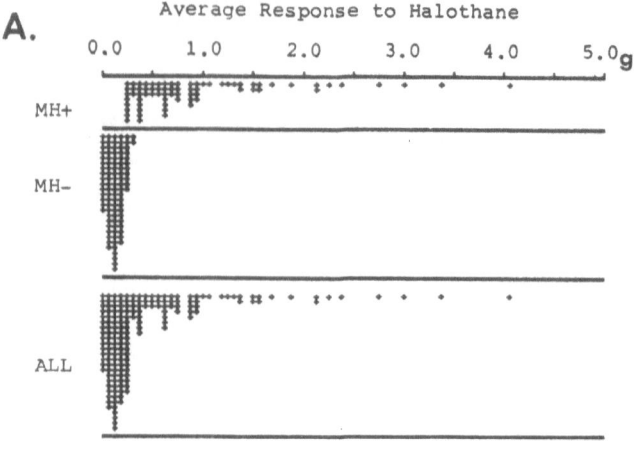

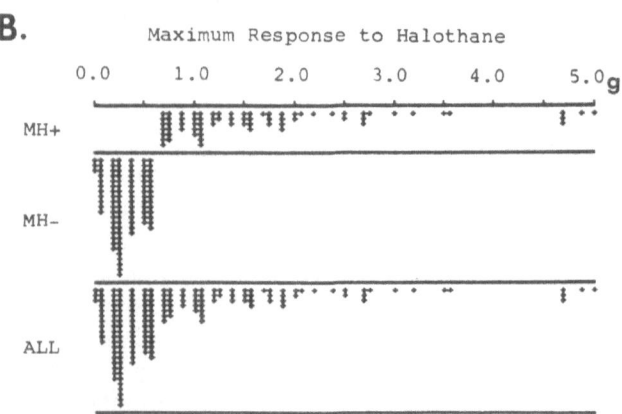

FIGURE 11.1. Frequency distribution of: (A) average response of all strips from each patient biopsy and; (B) the strip with the maximum response to halothane for each patient. Each "+" denotes one patient.

patients, one with a positive response to halothane and one with a negative response, are not clearly distinguishable. Although a bell–shaped curve is evident in the lower end of the response region associated with negative responders, the positive responses are not clustered about a single higher value (Figure 11.1). Rather the responses of MHS patients continue outward from the negative region. For diagnostic purposes, we believe, based on clinical correlations, that any abnormal response (i.e., greater than or equal to 0.7 g) to halothane 3% in the gas phase, even if only observed in a single fiber bundle from a biopsy specimen, should be treated as indicating MH susceptibility. This is a compromise between preventing false negatives (the major concern) and minimizing false positives.

In general, most of the muscle strips isolated from a single biopsy respond similarly to halothane. Clearly, the most troublesome problem in contracture testing for MH is the occasional variation between the responses of fascicles isolated from the same muscle biopsy of some patients (Figure 11.2). Britt et al.[26] observed positive contractures in only 13 of 30 (43%) fascicles from MH+ patients. We previously reported a similar percentage (54%) of positively responding strips from MHS patients.[32] In the latter study, less than 25% of the strips tested positive in the biopsies from 11 of 39 MHS patients. Nelson et al.[28] have demonstrated that as few as one of four fiber bundles from a known MHS pig can exhibit a contracture response to halothane. However, the weak in vitro response in muscle from this pig was paralleled by a lesser in vivo response to halothane as compared to the pigs with more consistent in vitro responses to halothane. Therefore, to prevent sampling errors, it is best to test six to eight strips for each patient. We regard those people in whom only one muscle strip from the biopsy has a positive test to halothane to be MH equivocal (MHE), however, for clinical purposes these patients are treated as MHS. The MHE diagnosis has also been found to be essential in the European classification system[30] and comprises about 15% of the total patients diagnosed.[31] The clinical significance of these nonhomogeneously responding biopsy specimens is difficult to interpret. These patients may develop MH only under prolonged surgery, or they may be more susceptible to MH at particular times (i.e., duration or type of anesthesia, age, season, disease state, presence of other drugs). This statement is pure speculation and remains to be tested.

Interpretation of Results

There are two proposed methods of interpreting the results obtained from halothane contracture testing. Method one requires a positive response from either the static halothane test or the caffeine contracture test for the diagnosis of MH+.[21] Method two requires a positive response to both the halothane and caffeine tests for the diagnosis of MHS.[30,31] In the first method, a minimum threshold for a contracture response at a particular

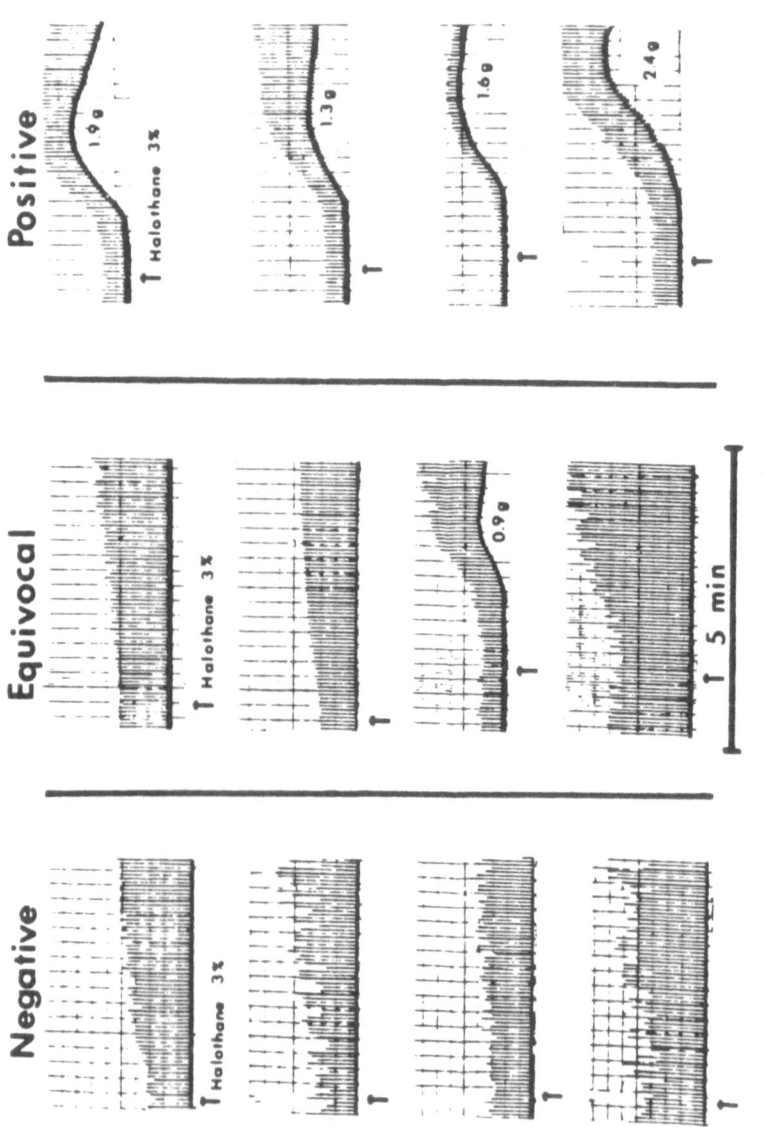

FIGURE 11.2. Spectrum of response to halothane contracture testing. Four fiber bundles were tested from biopsies of vastus lateralis of three patients. Notice that for some patients, not all the strips respond in the same manner.

concentration of halothane was established. For example, we have required a contracture of greater than or equal to 0.5 g to halothane 1% to 2%[21] or greater than or equal to 0.7 g to halothane 3%[12] for diagnosis of MH +. This latter criterion does not differ markedly from the former in which true control patients were tested. Alternatively, a contracture > 0.3 g to caffeine 2 mM would result in the diagnosis of MH +.[21]

The second method is to interpret a contracture threshold (0.2 g) to halothane 2% or less as a positive MH response.[30,31] This method would produce a large number of false positives, based on our experience.[21] The European Malignant Hyperpyrexia Group has reduced the introduction of false positives by also requiring a contracture threshold (0.2 g) occurring at 1.5 mM[30] or 2 mM[31] caffeine.

Possible Additional Phenotypes

The genetics of MH in humans are not fully understood. At least two phenotypes are readily agreed upon, that is, the MH − and MH + phenotypes. The Britt group[26] has reported that the HCSC (concentration of caffeine causing a 1 g contracture in the presence of halothane) was the most reliable means of differentiating MHS from normal patients. It was well established that the HCSC for the MH + population was abnormally low.[8,25,26,33] However, there are patients whose skeletal muscle does not respond to halothane alone and yet is abnormally sensitive in the HCSC test. Nelson et al.[25] have conducted an extensive investigation of MHS pigs in which the in vitro response to halothane was negligible and the HCSC was in the range of MHS. This group was referred to as phenotype K. The first recognition of a third phenotype was reported by others.[15] The in vivo challenge of phenotype-K pigs with halothane did not elicit the characteristic signs of MH.[25] A combination of succinylcholine and halothane in vivo did result in some of the signs of MH in this phenotype-K group of pigs; however, this did not appear to be a complete MH syndrome.

The clinical significance of the phenotype-K is unclear in humans. Nine of 23 patients with a history of anesthesia and diagnosed as phenotype K had manifested at least some signs suggestive of MH during anesthesia.[25] What was not understood was whether the phenotype-K patient had the potential for a full MH episode under extreme circumstances, or if this population would just exhibit some signs of MH. Patients in the phenotype-K group have the most mild clinical reactions.[2,26] Perhaps the mechanisms underlying this phenotype-K are different from those in true MH, or, as suggested by Nelson et al.,[25] the phenotype-K is one of a spectrum of contracture phenotypes for MH. As mentioned previously, the occurrence of a low HCSC in the normal population greatly exceeds the reported incidence of MH.

Contracture Tests: Methodology

The Biopsy

The muscle used in these tests was most often the vastus lateralis. However, the rectus abdominus muscle was considered a good alternative.
Muscle biopsies can safely be done under femoral block, using mepivacaine
or lidocaine,[34] although some investigators prefer general anesthesia.[29,35]
The size of the biopsy specimen was approximately 2.5 × 3.0 × 0.3 cm
and usually was obtained on a special clamp.[36] We usually also test a free
unclamped piece of muscle. Dimensions of the muscle strips isolated from
the biopsy and used for mounting in the tissue baths are generally 1.0 to
2.0 cm (length) by 2 to 5 mm (width) by 1 to 4 mm (thickness).

Halothane Contracture Test: Static

Muscle strips are mounted in a tissue bath containing a modified Krebs
solution (see reference 21 for composition) at 37°C and bubbled with
O_2:CO_2 (95%:5%) and the resting tension is usually adjusted to 2.0 g (Figure

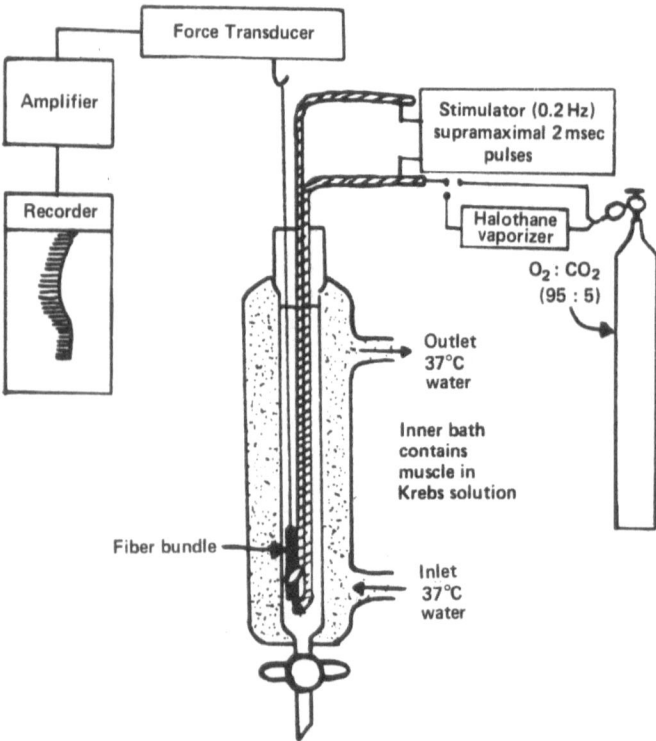

FIGURE 11.3. Diagrammatic representation of the in vitro bath used for muscle
contracture testing.

11.3). The bundles are directly stimulated supramaximally (10 V to 30 V) with pulses of 2 to 10 msec duration at a frequency of 0.2 Hz. Following an equilibration period (usually 10 to 30 minutes), halothane in 95% O_2:5% CO_2 is bubbled through the chamber. In our laboratory, patients are judged MHS if contractures in any one of the six to eight strips tested are equal to or greater than 0.7 g within 5 minutes of exposure to halothane 3%[12] or greater than 0.5 g after exposure to halothane 1% to 2%.[21] Since the contracture response of muscle preparations can vary somewhat from laboratory to laboratory and depend on the specific muscle examined, it was necessary for each laboratory to establish its own criteria for susceptibility based on studies with known control and MHS patients.

Halothane Contracture Test: Dynamic

The dynamic halothane contracture test was one of three tests used in the European standard for MH testing.[30] The test was first introduced as a means to detect susceptibility to MH of the nonrigid form.[10] The primary difference between the static and dynamic halothane contracture tests was the manner in which muscle tension was maintained. In the static test, a constant resting tension was used. In the dynamic test, the muscle was allowed to rest at a fixed tension for several minutes and then was stretched at a rate of 4 mm/minute for 1.5 minutes. The preparation was held at this length for 1 minute before decreasing the tension at the same rate. After three control cycles, halothane 0.5% was added for 3 minutes, and another cycle was completed. The halothane concentration was then increased (1.0, 2.0 and 3.0%) with each subsequent cycle.[22,30] A threshold of contracture of 0.2 g was considered a positive response. A recent study suggests that the dynamic test may be more sensitive than the static test in detecting MHS patients.[22] We believe that similar increased sensitivity can be achieved with the static halothane test by testing at least six muscle strips from a single biopsy specimen.[32]

Caffeine Contracture Test and Caffeine–Specific Concentration (CSC)

The muscle strips are mounted in the manner described previously for the static halothane contracture test. Following an equilibration in Krebs solution bubbled with O_2/CO_2 (95%:5%), strips are exposed to concentrations (0.25 to 16 mM) of caffeine increasing at twofold intervals. Caffeine was injected into the bath from a concentrated solution. Several diagnostic criteria have been used, with two having been agreed upon. The most common was the caffeine-specific concentration (CSC) in which the concentration of caffeine causing a 1 g contracture in the absence of halothane was estimated. If the CSC was less than about 4.0 mM, the patient was diagnosed as MH + .[29] The second criterion is a contracture greater than or equal to 0.3 g to caffeine at 2 mM. This criterion has been proposed as a standard base of comparison among laboratories.[21]

Halothane–Caffeine Specific Concentration (HCSC) Test

The halothane–caffeine contracture test is a variation of the caffeine contracture test described previously. In the halothane–caffeine contracture test, halothane at a concentration of 1% is bubbled through the bath during the caffeine dose–response. The HCSC, or that concentration of caffeine causing a 1 g contracture, was estimated from the log dose–response curve. If the HCSC was less than about 1.0 mM, then the patient was diagnosed as MH+.[29]

Contracture Tests in Stages of Development

Succinylcholine in the absence of any other agent does not induce contractures in skeletal muscle from MHS or normal people.[12–15] We reported that succinylcholine acts synergistically with halothane,[12] but we have not fully examined this interaction. Similar results have recently been reported by others,[37] who found that muscle from phenotype K patients had abnormally large contractures to halothane following succinylcholine exposure. We have since found that the order of agent addition is very important in this synergism.[16,17,38]

Although the contracture response to succinylcholine following halothane exposure was statistically different between the MH− and MH+ groups (Table 11.1), this test does not appear to be a specific probe for diagnosis of MH. We do not recommend this test, even in combination with other tests, for diagnostic purposes. The contracture response to succinylcholine in the presence of halothane may be of use in screening drugs for antagonizing succinylcholine-induced MH, as previously re-

TABLE 11.1. Interaction between halothane and succinylcholine

Order of addition				
First	Second	MH−	MH+	Significance
		Contracture in g		(*t*-test)
Halothane	SCh	0.57 ± 0.05(39)	1.04 ± 0.07(36)	$P<0.001$
SCh	Halothane	0.58 ± 0.06(31)	1.17 ± 0.10(33)	$P<0.001$
First	Second	MH−/MMR−	MH−/MMR+	
		Contracture in g		
Halothane	SCh	0.59 ± 0.05(44)	0.64 ± 0.06(26)	NS
SCh	Halothane	0.54 ± 0.05(44)	0.73 ± 0.08(26)	$P<0.05$

SCh, succinylcholine; MMR, masseter muscle rigidity. The average response for the combined strips from a single biopsy is shown as the mean ± SEM (n). The first agent was added to the bath 5 min before the response to the second agent was recorded. For each muscle strip the maximum response to the second agent within 5 min was used.

ported.[16,17,38] This test does correlate significantly ($P < 0.05$) with the HCSC for MH − and MH + patients (unpublished observations).

The strongest responders to halothane following succinylcholine exposure among the MH − patients appear to be those patients with a history of masseter muscle rigidity (MMR)[12] (Table 11.1). Therefore, this test does not appear to distinguish between MH + patients and MH − patients with a history of MMR. It should be noted that the MH − /MMR − group may have included patients predisposed to MMR, but with no clinical history of MMR because they were not exposed to triggering agents. The results from such patients would artificially elevate the MH − /MMR − values. Although the diagnostic value of the response to halothane following succinylcholine exposure remains unclear, the significant correlation ($P < 0.05$) between this test and the halothane contracture test (unpublished observations) suggests at least some use once this response is better understood. At this time we would suggest that the response to halothane following succinylcholine exposure should not be used for diagnosis of MH.

Halsall and Ellis[14] have reported that succinlycholine acts synergistically with caffeine to induce contractures and that this test could be used for screening MH susceptibility. It was suggested that this test might be used to distinguish various phenotypes. However, the succinylcholine-caffeine test was not included in the European Malignant Hyperthermia Group's testing protocol.[30,31]

A recent study in pigs suggested that the Ca^{2+} ionophore A23187 could be used to distinguish those pigs susceptible to MH.[39] The effects of A23187 in a range of 1 to 4 µg/ml on MHS muscle and control muscle differed in two ways. First, A23187 increased twitch tension in normal intercostal fiber bundles (tendon to tendon), whereas, it decreased twitch tension in muscle from MHS pigs. Second, the magnitude of contractures induced by A23187 in muscle from MHS pigs was significantly greater than that of controls.

Important Factors in Contracture Testing

Most laboratories express the contracture results in terms of grams, without regard to fascicle size. Gallant et al.[40] suggest that a better method is to express tension with respect to cross-sectional area. We have not observed any relationship between fascicle size and magnitude of contracture response (unpublished observations), in agreement with others.[15,26] Therefore, we do not believe there is a need to express the contracture in terms of cross-sectional area.

Halothane-induced contractures have been reported as temperature dependent[28] and temperature independent.[26] Our own experience supports the former conclusion. We have observed that halothane bubbling through the bath at 22°C did not elicit contractures in the two preparations tested. Raising the temperature to 37°C for the same preparations, with halothane still bubbling through the bath, resulted in a marked contracture. Therefore, we believe the bath temperature must be maintained at 37°C during halothane testing. Caffeine contractures also appear to be temperature dependent, being greater at 37°C than at 22°C.[26,27] Oddly, the HCSC does not appear to be temperature sensitive.[26,28]

There are also temperature considerations with regard to transportation of the biopsy specimens to the laboratory. Muscle kept at 4°C is less reactive in subsequent diagnostic testing than that kept at 22°C or 37°C.[2] Muscle kept at room temperature may exhibit a spontaneous contracture when introduced into the bath.[11] These contractures do not appear if preparations are maintained at 37°C before mounting in the tissue bath.[2] Muscle kept at 37°C before testing fatigues more rapidly than that kept at 22°C.[2] Therefore, we recommend that preparations be maintained at room temperature before being mounted in the tissue bath.

Preparations should be tested for viability during contracture testing. Direct stimulation of the muscle (10 V to 30 V of 20 msec duration at 0.2 Hz) should result in a twitch height of at least 0.5 g. Most preparations should exhibit considerably greater twitch heights. In addition, many laboratories are equilibrating preparations for 20 to 30 minutes before testing.[12,22] We require a stable resting tension during this period. Our preparations remain viable for 3 to 4 hours. The height of the directly stimulated muscle twitch does not appear to correlate with the strength of contracture (unpublished observations).

There is general agreement that the halothane concentration should be sampled in the gas phase, since the calibration marks on the vaporizers are not always reliable. Additionally, laboratories should determine the halothane concentration in the bathing medium. There is general agreement that fresh strips should be used for each test. Britt et al.[26] report that, if the number of viable strips is low, a caffeine test can be performed without greatly affecting subsequent testing on the same fascicle.

Some investigators use curare in the bath at a concentration of about 1 μM in order to exclude acetylcholine receptor influences. We have not found the inclusion of curare to influence significantly the contracture response to halothane (unpublished observations), in agreement with Kalow et al.[15] We do not include it in our bathing medium.

Calcium Uptake Test

Very small muscle samples (100 mg) can be analyzed using the calcium uptake test.[41] However, a muscle biopsy still has to be performed. The primary advantage of this approach is that samples can be frozen in liquid N_2 and shipped to a laboratory capable of performing the test. This test, as applied to MH, has not been published in manuscript form. One published report is in abstract form[18] and another is a letter.[42] In brief, Ca^{2+} uptake into cryostat sections and actomyosin ATPase activity appear to be reduced in muscle from MH patients. A recent study has failed to confirm the validity of the Ca^{2+} uptake test.[43] In addition, a double-blind study of 28 patients, in whom the diagnosis by halothane contracture test was compared to diagnosis by the Ca^{2+} uptake test, has shown a large number of false positives by the Ca^{2+} uptake test (Gronert, personal communications).

Skinned Fiber Test

A major advantage of the skinned fiber test is that, as with the Ca^{2+} uptake test, samples can be shipped to other laboratories. The Ca^{2+}-induced Ca^{2+} release (Ca-ICaR) process appears to have an increased sensitivity to Ca^{2+} in muscle from MHS patients using skinned fiber preparations.[44] In addition, preparations from MHS patients have a reduced threshold to caffeine-induced Ca^{2+} release.[33,45,46] A good correlation has been established between caffeine-induced Ca^{2+} release and the HCSC.[33] However, the results of the HCSC test are not accepted as evidence of MH susceptibility by all laboratories.

Abnormal Skeletal Muscle Proteins

A study using sodium dodecyl sulfate (SDS) gel electrophoresis has suggested the presence of abnormal proteins in skeletal muscle from MHS patients.[47] This would appear to be a relatively simple method of MH diagnosis. However, these proteins were later found in control as well as MHS patients.[48,49] Blanck[50] addressed this discrepancy and confirmed that the appearance of the abnormal proteins was related to hemoglobin, which was absent in the control specimens due to a slight difference in technique used for biopsy. A highly detailed study of muscle proteins, including Ca^{2+} binding proteins, has shown no difference between MH− and MH+ patients.[51] Other investigators have also found no obvious difference between MHS and normal humans in protein profiles of skeletal muscle.[52–54]

Electromyography

A potential test using electromyography has been reported.[55] In this noninvasive test, the effects of succinylcholine on the electromyographic recordings of MHS patients differed from those observed for controls.

Nuclear Magnetic Resonance (NMR) Spectroscopy

Nuclear magnetic resonance (NMR) spectroscopy can examine changes in high energy phosphates. Preliminary studies in pigs undergoing an MH episode demonstrate extreme uncoupling in oxidative phosphorylation.[56] Some modification of this noninvasive approach might be informative in humans. The problem is how to cause the changes in high energy phosphate profiles for diagnostic purposes without triggering an MH episode in the human patient.

Intracellular Ca^{2+} Concentrations in Intercostal Muscle

Using Ca^{2+} selective microelectrodes Lopez et al.[57] have demonstrated elevated free myoplasmic Ca^{2+} in intercostal muscle removed from MHS patients. This test requires a muscle biopsy and an even more elaborate approach than the standard contracture test.

Tests Using Blood Components

The major problem in developing noninvasive diagnostic tests for MH susceptibility using blood components is that little is known about the mechanisms underlying MH. The extent to which the defect is expressed in tissue other than skeletal muscle is unknown.

Red Blood Cells

Several studies have demonstrated increased fragility in red blood cells from MHS pigs.[58-62] Reports on red blood cell fragility in humans have been less consistent. The fragility of red blood cells from MHS humans has been shown to be increased,[63-65] decreased,[66] or the same [63,66,67] relative to controls. We have used a slight variation of the method described by Dacie and Lewis,[68] in which we exclude PO_4 buffer during incubation (our pH = 7.0) and incubate the cells at 37°C. We have not observed a significant difference between the mean corpuscular fragility (MCF) of red blood cells from four MH− (0.432 ± 0.011, mean ± SEM) and five MH+ (0.428 ± 0.019) patients, as diagnosed by halothane and caffeine contracture testing. Refrigerating the cells overnight to enhance fragility did not significantly alter the MCF for either the MH− (0.441 ± 0.007) or MH+ (0.432 ± 0.012) group (unpublished observations).

Platelets (Nucleotide Depletion Test)

The use of platelet model systems for the diagnosis of MH was reviewed by Ellis and Heffron.[3] The platelet nucleotide depletion test was recently published.[19] In brief, platelets are incubated with halothane and the decrease in the energy charge ratio (ATP + ADP/hypoxanthine + AMP) is monitored using HPLC. Platelets from patients susceptible to MH were reported to have a more greatly reduced energy charge ratio than do those from normal patients. Several other studies have failed to validate the nucleotide depletion test.[69-72]

White Blood Cells

In a recent review, Britt[29] briefly described a test being developed that is based on lymphocytes. Halothane causes a greater increase in cytoplasmic Ca^{2+}, as determined by fluorescent indicators, for preparations from MH+ than MH− patients.

Serum Creatine Kinase (CK)

High resting CK values were once believed to be a useful diagnostic tool for MH.[73] Despite a tendency for resting CK values to be elevated in some MH patients, subsequent studies have not found resting CK values to be a reliable indicator of MH susceptibility,[74-77] even in family members of MHS patients. There is a high incidence of MH in patients exhibiting MMR during anesthesia.[32] We have conducted a retrospective survey of patients referred to our laboratory for diagnosis after experiencing an episode of MMR. In this study, we calculated the probability of a positive contracture test with the postoperative CK values. The results of this study are shown in Table 11.2. In the absence of an underlying myopathy, a postoperative CK value >20,000 is a good indicator of MH susceptibility.[32,78]

Discussion

The outcome of the halothane contracture test in combination with that of the response to caffeine following one of the established protocols,[21,30,31] or a slight modification[12] is currently the method of choice for diagnosis of MH. A large number of fiber bundles (six to eight) should be tested with halothane for maximum sensitivity. An additional one or two strips should also be tested with caffeine. All other tests should be considered

TABLE 11.2. Predictability of malignant hyperthermia susceptibility based on postoperative CK values of patients exhibiting MMR

CK Value	Probability of MH+ Diagnosis by contracture test
>20,000	100%
>15,000	88%
>10,000	82%
> 7,500	66%

The patients studied comprised 21 MH negative patients and 30 MH positive patients.

either preliminary or not widely accepted. However, much remains to be done to confirm the validity of contracture tests and develop less invasive means of diagnosis. A better understanding of the mechanisms underlying MH would greatly aid the area of diagnosis.

References

1. Britt BA (1979) Etiology and pathophysiology of malignant hyperthermia. Fed Proc 38:44–48
2. Britt BA (1982) Malignant hyperthermia: A review, in Milton AS (ed) *Handbook of Pharmacology*. Berlin: Springer-Verlag, vol 60, pp 547–615
3. Ellis FR, Heffron JJA (1985) Clinical and biochemical aspects of malignant hyperpyrexia, in Atkinson RS, Adams AP (eds) *Recent Advances in Anaesthesia and Analgesia*. Edinburgh: Churchill Livingstone, pp 173–207
4. Gronert, GA (1980) Malignant hyperthermia. Anesthesiology 53:395–423
5. Rosenberg H, Fletcher JE (1986) Malignant hyperthermia, in Barash-PG (ed) *ASA Refresher Courses in Anesthesiology*. Philadelphia: JB Lippincott Co, vol 14, pp 207–216
6. Rosenberg H, Fletcher JE (1987) Malignant hyperthermia, in Azar I (ed) *Muscle Relaxants: Side Effects and a Rational Approach to Selection*. New York: Marcel Dekker, pp 115–148
7. Gronert GA (1983) Controversies in malignant hyperthermia. Anesthesiology 59:273–274
8. Kalow W, Britt BA, Terreau ME, Haist C (1970) Metabolic error of muscle metabolism after recovery from malignant hyperthermia. Lancet 2:895–898
9. Ellis FR, Harriman DGF, Keaney NP, Kyei-Mensah K, Tyrrell JH (1971) Halothane-induced muscle contracture as a cause of hyperpyrexia. Br J Anaesth 43:721–722
10. Ellis FR, Harriman DGF (1973) A new screening test for susceptibility to malignant hyperpyrexia. Br J Anaesth 45:638
11. Moulds RFW, Denborough MA (1974) Biochemical basis of malignant hyperpyrexia. Br Med J 2:241–244.
12. Fletcher JE, Rosenberg H (1985) *In vitro* interaction between halothane and succinylcholine in human skeletal muscle: Implications for malignant hyperthermia and masseter muscle rigidity. Anesthesiology 63:190–194
13. Galloway GJ, Denborough MA (1986) Suxamethonium chloride and malignant hyperpyrexia. Br J Anaesth 58:447–450
14. Halsall PJ, Ellis FR (1979) A screening test for the malignant hyperpyrexia phenotype using suxamethonium-induced contracture of muscle treated with caffeine, and its inhibition by dantrolene. Br J Anaesth 51:753–756
15. Kalow W, Britt BA, Richter A (1977) The caffeine test of isolated human muscle in relation to malignant hyperthermia. Can Anaesth Soc J 24:678–694
16. Fletcher JE, Rosenberg H (1986) In vitro muscle contractures induced by halothane and suxamethonium. I. The rat diaphragm. Br J Anaesth 58:1427–1432
17. Fletcher JE, Rosenberg H (1986) In vitro muscle contractures induced by halothane and suxamethonium. II. Human skeletal muscle from normal and malignant hyperthermia susceptible patients. Br J Anaesth 58:1433–1439

18. Allen PD, Ryan JF, Sreter FA, Mabuchi K (1980) Rigid vs non rigid MH, studies of Ca^{2+} uptake and actomyosin ATPase (Abstr). Anesthesiology 53:S251

19. Solomons CC, Masson NC (1984) Platelet model for halothane-induced effects on nucleotide metabolism applied to malignant hyperthermia. Acta Anaesth Scand 28:185–190

20. Ørding H, Ranklev E, Fletcher R (1984) Investigation of malignant hyperthermia in Denmark and Sweden. Br J Anaesth 56:1183–1190

21. Rosenberg H, Reed S (1983) In vitro contracture tests for susceptibility to malignant hyperthermia. Anesth Analg 62:415–420

22. Ranklev E, Fletcher R, Blomquist S (1986) Static vs Dynamic tests in the in vitro diagnosis of malignant hyperthermia susceptibility. Br J Anaesth 58:646–648

23. Ellis FR, Harriman DGF, Currie S, Cain PA (1978) Screening for malignant hyperthermia in susceptible patients, in Aldrete JA, Britt-BA (eds) *The Second International Symposium on Malignant Hyperthermia*. New York: Grune & Stratton, pp 273–285

24. Rosenberg, H (1981) International workshop on malignant hyperpyrexia. Anesthesiology 54:530–531

25. Nelson TE, Flewellen EH, Gloyna DF (1983) Spectrum of susceptibility to malignant hyperthermia - Diagnostic dilemma. Anesth Analg 62:545–552

26. Britt BA, Endrenyi L, Scott E, Frodis W (1980) Effects of temperature, time and fascicle size on the caffeine contracture test. Can Anaesth Soc J 27:1–11

27. Nelson TE, Austin KL, Denborough MA (1977) Screening for malignant hyperpyrexia. Br J Anaesth 49:169–172

28. Nelson TE, Bedell DM, Jones EW (1975) Porcine malignant hyperthermia: Effects of temperature and extracellular calcium concentration on halothane-induced contracture of susceptible skeletal muscle. Anesthesiology 42:301–306

29. Britt BA (1985) Malignant hyperthermia. Can Anaesth Soc J 32:666–677

30. European Malignant Hyperthermia Group (1984) A protocol for the investigation of malignant hyperpyrexia (MH) susceptibility. Br J Anaesth 56:1267–1269

31. European Malignant Hyperthermia Group (1985) Laboratory diagnosis of malignant hyperpyrexia susceptibility (MHS). Br J Anaesth 57:1038–1046

32. Rosenberg H, Fletcher JE (1986) Masseter muscle rigidity and malignant hyperthermia susceptibility. Anesth Analg 65:161–164

33. Britt BA, Frodis W, Scott E, Clements MJ, Endrenyi L (1982) Comparison of the caffeine skinned fibre tension (CSFT) test with the caffeine-halothane contracture (CHC) test in the diagnosis of malignant hyperthermia. Can Anaesth Soc J 29:550–562

34. Berkowitz A, Rosenberg H (1985) Femoral block with mepivacaine for muscle biopsy in malignant hyperthermia patients. Anesthesiology 62:651–652

35. Britt BA (1986) Malignant hyperthermia: Choice of anaesthesia for muscle biopsy (Letter-Reply). Can Anaesth Soc J 33:419–420

36. Berman AT, Garbarino JL, Rosenberg H, Heiman- Patterson T, Bosacco SJ, Weiss AA (1985) Muscle biopsy: Proper surgical technique. Clin Orthop 188:240–243

37. Kennamer DL, Belt MW, Winsett O, Nelson TE (1986) A comparison of suc-

cinylcholine contracture effects on malignant hyperthermia diagnostic human muscle (Abstr). Anesthesiology 65:A237

38. Fletcher JE, Rosenberg H, Hilf M (1985) In vitro studies of droperidol for use in human malignant hyperthermia (Abstr). Anesthesiology 63:A302

39. Reiss G, Monin G, Lauer C (1986) Comparative effects of the ionophore A23187 on the mechanical responses of muscle in normal Pietrain pigs and pigs with malignant hyperthermia. Can J Physiol Pharmacol 64:248–253

40. Gallant, EM, Fletcher TF, Goettl VM, Rempel WE (1986) Porcine malignant hyperthermia: Cell injury enhances halothane sensitivity of biopsies. Muscle Nerve 9:174–184

41. Mabuchi K, Sreter FA (1978) Use of cryostat sections for measurement of Ca^{2+} uptake by sarcoplasmic reticulum. Analyt Biochem 86:733–742

42. Allen PD, Ryan JF, Jones DE, Mabuchi K, Virga A, Roberts J, Sreter F (1986) Sarcoplasmic reticulum calcium uptake in cryostat sections of skeletal muscle from malignant hyperthermia patients and controls. Muscle Nerve 9: 474–475

43. Nagarajan K, Fishbein WN, Carlin HM, Pezeshkpour G, Muldoon SM (1985) Frozen section calcium uptake versus halothane and caffeine contracture tests on human muscle (Abstr). Anesthesiology 63:A307

44. Endo M, Yagi S, Ishizuka T, Horiuti K, Koga Y, Amaha K (1983) Changes in the Ca-induced Ca release mechanism in the sarcoplasmic reticulum of the muscle from a patient with malignant hyperthermia. Biomed Res 4:83–92

45. Araki M, Takagi A, Fujita T, Matsubara T (1985) Porcine malignant hyperthermia: Caffeine contracture of single skinned muscle fibers. Biomed Res 6:73–78

46. Takagi A, Sunohara N, Ishihara T, Nonaka I, Sugita H (1983) Malignant hyperthermia and related neuromuscular diseases: Caffeine contracture of the skinned muscle fibers. Muscle Nerve 6:510–514

47. Blanck TJJ, Fisher YI, Thompson M, Muldoon S (1984) Low molecular weight proteins in human malignant hyperthermia muscle. Anesthesiology 61:589–592

48. Fletcher JE, Rosenberg H (1985) Low molecular weight proteins in human malignant hyperthermia muscle (Letter). Anesthesiology 62:849–850

49. Fletcher JE, Rosenberg H, Hilf M (1984) Electrophoresis of soluble muscle protein from human malignant hyperthermia susceptibles (Abstr). Anesthesiology 61:A279

50. Blanck TJJ (1985) Low molecular weight proteins in human malignant hyperthermia muscle (Letter-Reply). Anesthesiology 62:850

51. Walsh MP, Brownell KW, Littman V, Paasuke RT (1986) Electrophoresis of muscle proteins is not a method for diagnosis of malignant hyperthermia susceptibility. Anesthesiology 64:473–479

52. Marjanen LA, Denborough MA (1984) Electrophoretic analysis of proteins in malignant hyperpyrexia susceptible skeletal muscle. Int J Biochem 16:919–929

53. Oku S, Liew C-C, Britt BA (1983) Analysis of sarcoplasmic reticulum proteins in patients susceptible to malignant hyperthermia. J Neurol Sci 60:127–135

54. Whistler T, Isaacs H, Badenhorst M (1986) No abnormal low molecular weight proteins identified in human malignant hyperthermic muscle. Anesthesiology 64:795–797

55. Eng GD, Becker MJ, Muldoon SM (1984) Electrodiagnostic tests in the detection of malignant hyperthermia. Muscle Nerve 7:618–625

56. Roberts JT, Burt T, Gouylai L, Chance B, Sreter F, Ryan J (1983) Immediate uncoupling of high energy oxidative phosphorylation in muscle of malignant hyperthermic swine determined non-invasively by whole body ^{31}P nuclear magnetic resonance (Abstr). Anesthesiology 59:A230

57. Lopez JR, Alamo L, Caputo C, Wikinski J, Ledezma D (1985) Intracellular ionized calcium concentration in muscles from humans with malignant hyperthermia. Muscle Nerve 8:355–358

58. Cheah KS, Cheah AM (1979) Mitochondrial calcium, erythrocyte fragility and porcine malignant hyperthermia. FEBS Lett 107:265–268

59. Harrison GG, Verburg C (1973) Erythrocyte osmotic fragility in hyperthermia-susceptible swine. Br J Anaesth 45:131–133

60. Heffron JJA, Mitchell G (1979) Influence of pH, temperature, halothane and its metabolites on osmotic fragility of eryhrocytes of malignant hyperthermia-susceptible and resistant pigs. Br J Anaesth 53:499–504

61. King WA, Ollivier L, Basrur PK (1976) Erythrocyte osmotic response test on malignant hyperthermia-susceptible pigs Ann Genet Sel Anim 8:537–540

62. O'Brien PJ, Rooney MT, Reik TR, Thatte HS, Rempel WE, Adiss PB, Louis CF (1985) Porcine malignant hyperthermia susceptibility: Erythrocyte osmotic fragility. Am J Vet Res 46:1451–1456

63. Ellis FR (1981) Malignant hyperpyrexia. Monogr Anaesth 9:163–169

64. Kelstrup J, Hasse J, Jorni J, Reske-Nielsen R, Hanel HK (1973) Malignant hyperthermia in a family. Acta Anaesth Scand 17:283–284

65. Shimonaka H, Yamamoto M, Tanahashi T et al. (1983) Biochemical studies on erythrocyte membranes of myoglobinuria. Hiroshima J Anesth 19:53–58

66. Godin DV, Herring FG, MacLeod PJM (1981) Malignant hyperthermia: Characterization of erythrocyte membranes from individuals at risk. J Med 12:35–49

67. Zsigmond EK, Penner J, Kothary SP (1978) Normal erythrocyte fragility and abnormal platelet aggregation in MH families: A pilot study, in Aldrete-JA, Britt BA (eds) The Second International Symposium on Malignant Hyperthermia. New York: Grune & Stratton, pp 213–219

68. Dacie JV, Lewis SM (1968) Practical Haematology, 4th ed. New York: Grune & Stratton

69. Britt BA, Scott EA (1986) Failure of the platelet-halothane nucleotide depletion test as a diagnostic or screening test for malignant hyperthermia. Anesth Analg 65:171–175

70. Giger U, Kaplan RF (1983) Halothane induced ATP depletion in platelets from patients susceptible to malignant hyperthermia. Anesthesiology 58:347–352

71. Lee MB, Adragna MG, Edwards L (1985) The use of a platelet nucleotide assay as a possible diagnostic test for malignant hyperthermia. Anesthesiology 63:311–315

72. Verburg MP, Van Bennekom CA, Oerlemansft JJJ, De Bruyn CHMM (1984) Malignant hyperthermia: Adenine incorporation and adenine metabolism in human platelets, influenced by halothane. Adv Exp Med Biol 165:442–446

73. Isaacs H, Barlow MB (1970) The genetic background to malignant hyperpyrexia revealed by serum creatine phosphokinase estimations in asymptomatic relatives. Br J Anaesth 42:1077–1084

74. Amaranath L, Lavin TJ, Trusso RA, Boutros AR (1983) Evaluation of creatine phosphokinase screening as a predictor of malignant hyperthermia. Br J Anaesth 55:531–533

75. Britt BA, Endrenyi L, Peters PL, Kwong FHF, Kadijevic L (1976) Screening of malignant hyperthermia–susceptible families by creatine phosphokinase measurement and other clinical investigations. Can Anaesth Soc J 23:263–284

76. Ellis FR, Clarke IMC, Modgill M, Currie S, Harriman DGF (1975) Evaluation of creatine phosphokinase in screening patients for malignant hyperpyrexia. Br Med J 3:511–513

77. Paasuke RT, Brownell KW (1986) Serum creatine kinase level as a screening test for susceptibility to malignant hyperthermia. JAMA 255:769–771

78. Larach MG, Rosenberg H, Larach DR, Broennle AM (1987) Prediction of malignant hyperthermia susceptibility by clinical signs. Anesthesiology 66:547–550

CHAPTER 12

Malignant Hyperthermia: A Review

Michael Denborough

Malignant hyperthermia (hyperpyrexia) (MH) was first described in 1960[1] in dramatic circumstances, when an injured 21-year-old man nearly died following anesthesia with halothane. He had given a history that 10 of his close relatives had died as a result of anesthesia with ether. Since then, MH has come to be recognized as an important cause of anesthetic induced death, and has been the subject of extensive investigation.

It is now known that MH occurs in people with an underlying muscle membrane abnormality and an inability to regulate the myoplasmic calcium concentration.[2-4] Three predisposing myopathies have been defined and triggers of MH other than anesthesia have been identified.[3-7] Nonanesthetic triggers of MH may lead to diverse clinical presentations of the syndrome.[3-7]

Skeletal Muscle and Malignant Hyperthermia

The evidence that MH develops in people who have an underlying muscle disease came in 1969.[8] Following clinical reports that muscle rigidity may occur in an MH reaction, serum creatinephosphokinase (CPK) levels were measured in a man who developed the anesthetic complication. The results showed a massive elevation of the serum CPK from 250 IU/ml (normal range 0 to 70 IU) to 20,500 in 24 hours, indicating that the anesthetic had induced severe muscle damage in the susceptible patient. Subsequent investigations revealed high serum levels in asymptomatic relatives of patients who had developed MH.[9-12]

Three myopathies that predispose to MH have now been defined. The most common (Evans myopathy) is dominantly inherited and usually subclinical, although there may be some wasting, particularly in the lower parts of the thighs.[8] The second myopathy is much less common and is probably inherited as a recessive trait. It occurs in children, usually boys, and is associated with a number of physical abnormalities.[13] The third myopathy which predisposes to MH is central-core disease[6] which is usually inherited as a dominant characteristic. Histochemical and electron microscopic examinations of muscle in affected people show striking nonstaining lesions extending along type 1 fibers, and there is often type 1 atrophy.

Clinical Presentation of Malignant Hyperthermia

The full-blown syndrome usually occurs in people who have received a halogenated anesthetic and succinylcholine. There is a rapid and sustained rise in body temperature, without shivering, either during the operation or in the recovery room, in the absence of any obvious cause, such as infection or hot and humid environmental conditions. Tachycardia, tachypnea, cyanosis, generalized muscle rigidity, and cardiac arrhythmias are common. Acidosis is an early finding.[16,25]

Nonanesthetic triggers of MH can lead to clinical presentations as diverse as heatstroke,[4] renal failure,[5] the neuroleptic malignant syndrome,[3] and sudden infant death.[7]

Malignant Hyperthermia in Animals

Susceptibility to MH is not confined to humans but occurs also in a number of animals, including pigs,[18–21] dogs,[22] cats,[23] and horses.[21] The pig has been studied most extensively, and the pharmacologic properties of MH swine muscle appear to be identical with MH human muscle.[12,15,17,21]

Breeding for heavily muscled swine has led to the development of two syndromes, the porcine stress syndrome (PSS) and the pale, soft, exudative pork (PSEP) syndrome. PSS ia an acute shocklike syndrome produced by stress,[20] such as transportation, sudden increases in ambient temperature, mild exercise, or fighting, whereas PSEP refers to poor meat quality found postmortem.[18] The breeds of swine in which MH has been described are those in which there is a high incidence of PSS and PSEP, and in swine it appears that MH, PSS, and PSEP are different manifestations of the same syndrome.[21,24]

Biochemistry of Malignant Hyperthermia

The availability of swine susceptible to MH has provided a means of investigating the early biochemical changes that occur during an episode of MH. The first major chemical change is the production of lactic acid, and this is accompanied by a simultaneous drop in blood pH. These changes precede the rise in body temperature. The production of lactic acid is accompanied by a rapid rise in blood $PaCO_2$ and a subsequent fall in blood bicarbonate.[25]

These biochemical findings led to the suggestion that the primary event in MH might be a rise in the calcium ion concentration in the myoplasm.[25] This would explain the three main clinical features of MH, which are muscular rigidity, metabolic acidosis, and a rise in body temperature. The rise in the calcium ion concentration in the myoplasm would mean that these calcium ions would combine with troponin, allowing actin and myosin filaments to interdigitate and thus cause muscular contraction. Calcium ions also activate phosphorylase, so that glycogen would be broken down to lactic acid, and this would explain the metabolic acidosis. The activation of phosphorylase would keep up the supply of fructose 1, 6-disphosphate for ATP production by glycolysis. Heat would be generated during the continued synthesis and utilization of ATP during glycolysis in both muscle and liver, and this would explain the rise in temperature that occurs in MH.

Making use of the fact that the strength of a skeletal muscle contracture is a function of the concentration of free calcium ions in the cytoplasm, the demonstration of increased contractility in MHS muscle to chemical agents has confirmed that elevation of myoplasmic calcium concentration is a primary event in the development of MH.[14]

Site of the Muscle Cell Abnormality in Malignant Hyperthermia

The precise site of the muscle membrane abnormality that predisposes to MH is not yet known. In vitro studies, using agents that act specifically on different parts of the muscle contractile mechanism, suggest that the abnormality in MHS swine muscle lies distal to the postjunctional membrane.[12,27] Two possible sites are involved, one being the sarcolemma and the T system, the other the excitation-contraction coupling mechanism, and the sarcoplasmic reticulum.

Studies of Ca^{2+} uptake by the sarcoplasmic reticulum isolated from MHS swine have led to no firm conclusions. Dantrolene sodium was found to have no effect on Ca^{2+} uptake by MH sarcoplasmic reticulum preparations,[28] but had an inhibitory effect on halothane–induced calcium release.[17]

Identification of Susceptibility to Malignant Hyperthermia

Although some people who are susceptible to MH have raised levels of serum CPK, this test can give both false positive and false negative results.

The only definitive way to identify susceptibility to MH that is available at present is to demonstrate enhanced contractility of skeletal muscle in vitro. This was first described in 1970 by using caffeine,[14] which has long been known to cause skeletal muscle contraction. Although halothane contractures were not observed in this investigation, later studies showed spontaneous contractures of MHS muscle when exposed to halothane.[9,11] The reasons for the differences in these results was that halothane-induced contractures in MHS muscle are temperature dependent. Depolarization of the muscle cell membrane by potassium chloride also causes an increased contracture in MHS muscle.[13]

Although the in vitro muscle contracture test is a precise and accurate means of identifying susceptibility of MH, it is an invasive procedure. Therefore it causes discomfort for the patient and also is limited in the extent to which it can be used. A noninvasive test is needed to identify susceptibility to MH, and there are indications that the technique of topical nuclear magnetic resonance (NMR) may provide the answer.[30,31]

The technique of ^{31}P NMR can be used to determine relative levels of ATP, phosphocreatine (PCr), and inorganic phosphate (P_i) in skeletal muscle and can thus be used to study metabolism in healthy and diseased muscle. When muscle from swine that are susceptible to MH is exposed to halothane and caffeine in vitro, both cause a change in the NMR spectra.[30] The phosphocreatine peak is decreased in area, whereas the inorganic phosphate resonance shows a comparable increase in area. Substituting this test for an in vivo technique using topical NMR, would provide a simple method of identifying susceptibility to MH.

Treatment

Dantrolene sodium, 1-([5-(nitrophenyl) furfurylidene] amino) hydantoin sodium was synthesized by Norwich-Eaton Pharmaceuticals as a possible antibiotic. Toxicity studies showed that dantrolene sodium induced flaccidity when injected into mice. Subsequent investigations showed that this effect was produced by inhibiting excitation-contraction coupling in skeletal muscle and thus lowering the myoplasmic calcium concentration, without affecting neuromuscular transmission or the electrical properties of the muscle.[27]

Because of its effects on excitation–contraction coupling in skeletal muscle, the role of dantrolene has been studied in MH, both in vivo and in investigations on MHS muscle in vitro.[2,17,32] Dantrolene sodium causes complete, immediate, and sustained relaxation of caffeine and halothane–induced contractures in vitro in MHS muscle.[33] In vivo dantrolene sodium has been shown to be very effective in treating MH, both in humans[32] and in swine.[2]

Summary

Malignant hyperthermia (hyperpyrexia) (MH) is a dangerous complicaton of anesthesia occurring in people with an underlying muscle disease. Three predisposing myopathies have been defined. An apparently identical muscle abnormality occurs in certain animals. The primary biochemical abnormality is an elevation in the calcium ion concentration in the myoplasm, and the defect appears to lie in the excitation-contraction coupling mechanism. The precise site and nature of the abnormality have not yet been defined. Several nonanesthetic triggers of MH are now recognized, and these can lead to a variety of clinical presentations including heatstroke, renal failure, the neuroleptic malignant syndrome, and sudden infant death. Susceptibility to MH can be identified by demonstrating increased contractility of skeletal muscle in vitro, but topical nuclear magnetic resonance should provide a noninvasive screening test. Dantrolene sodium, a drug that lowers myoplasmic calcium ion concentration, is a specific and effective treatment for MH.

References

1. Denborough MA, Lovell RRH (1960) Anaesthetic deaths in a family. Lancet 2:45
2. Harrison GG (1975) Control of the malignant hyperthermia syndrome in MHS swine by dantrolene sodium. Br J Anaesth 47:62–65
3. Caroff S, Rosenberg H, Gerber JC (1983) Neuroleptic malignant syndrome and malignant hyperthermia. Lancet 1:244
4. Denborough MA (1982) Heat-stroke and malignant hyperpyrexia. Med J Aust 1:649–650
5. Denborough MA, Collins SP, Hopkinson KC (1984) Rhabdomyolysis and malignant hyperpyrexia. Br Med J 288:1878
6. Denborough MA, Dennett X, Anderson RMcD (1973) Central-core disease and malignant hyperpyrexia. Br Med J 1:272–273
7. Denborough MA, Galloway GF, Hopkinson KC (1982) Malignant hyperpyrexia and sudden infant death. Lancet 2:1068–1069
8. Denborough MA, Ebeling P, King JO, Zapf P (1970) Myopathy and malignant hyperpyrexia. Lancet 1:1138–1140
9. Ellis FR, Keaney NP, Harriman DGF, Summer DW, Kyei-Mensah K, Tyrrell JH, Hargreaves JB, Parikh RK, Mulrooney PL (1972) Screening for malignant hyperpyrexia. Br Med J 4:526–528
10. Isaacs H, Barlow MB (1970) The genetic background to malignant hyperpyrexia revealed by serum creatine phosphokinase estimations in asymptomatic relatives. Br J Anaesth 42:1007–1084
11. Moulds RFW, Denborough MA (1974) Identification of susceptibility to malignant hyperpyrexia. Br Med J 2:245–247
12. Okumura F, Crocker BD, Denborough MA (1979) Identification of susceptibility to malignant hyperpyrexia in swine. Br J Anaesth 51:171–176

13. Denborough MA (1984) Malignant hyperpyrexia. Clin Anesth 2:669–675
14. Kalow W, Britt BA, Terreau ME, Haist C (1970) Metabolic error of muscle metabolism after recovery from malignant hyperthermia. Lancet 2:895–898
15. Moulds RFW, Denborough MA (1974) Biochemical basis of malignant hyperpyrexia. Br Med J 2:241–244
16. Harrison GG, Biebuyck JF, Terblanche J, Dent DM, Hickman R, Saunders SJ (1968) Hyperpyrexia during anaesthesia. Br Med J 3:594–595
17. Ohnishi ST, Taylor S, Gronert GA (1983) Calcium-induced Ca^{2+} release from sarcoplasmic reticulum of pigs susceptible to malignant hyperthermia. The effects of halothane and dantrolene. FEBS Letts 162:103–107
18. Briskey EJ, Wiemer-Pedersen J (1961) Biochemistry of pork musculature. I. Rate of anaerobic glycolysis and temperature changes versus the apparent structure of muscle tissue. Food Sci 26:297–305
19. Hall LW, Woolf N, Bradley JWP, Jolly DW (1966) Unusual reaction to suxamethonium chloride. Br Med J 2:1305
20. Topel DG, Bicknell EJ, Preston KS, Christian LL, Matsushima CY (1968) Porcine stress syndrome. Mod Vet Prac 49:40–60
21. Williams CH (1976) Some observations on the etiology of the fulminant hyperthermia-stress syndrome. Perspect Biol Med 20:120–130
22. Short CE, Paddleford RR (1973) Malignant hyperthermia in the dog. Anesthesiology 39:462–463
23. De Jong RH, Heavner JE, Amory DW (1974) Malignant hyperthermia in the cat. Anesthesiology 41:608–609
24. Nelson TE (1971) Porcine stress syndromes, in Gordon RA, Britt BA, Kalow W (eds) *International Symposium on Malignant Hyperthermia*. Springfield, IL: Charles C. Thomas, pp 191–197
25. Denborough MA, Hird FJR, King JO, Marginson MA, Mitchelson KR, Naylor WG, Rex MA, Zapf P, Condron RJ (1971) Mitochondrial and other studies in Australian Landrace pigs affected with malignant hyperthermia, in Gordon RA, Britt BA, Kalow W (eds) *International Symposium on Malignant Hyperthermia*. Springfield, IL: Charles C Thomas, pp 229–237
26. Denborough MA, Forster JFA, Hudson MC, Carter NG, Zapf P (1970) Biochemical changes in malignant hyperpyrexia. Lancet 1:1136–1138
27. Ellis KO, Bryant SH (1972) Excitation-contraction uncoupling in skeletal muscle by dantrolene sodium. Naunyn-Schmiedebergs Arch Exp Path Pharmak 274:107–109
28. White MD, Collins JG, Denborough MA (1983) The effect of dantrolene on skeletal muscle sarcoplasmic reticulum function in malignant hyperpyrexia in pigs. Biochem J 212:399–405
29. Gronert GA (1980) Malignant hyperthermia. Anesthesiology 53:395–423
30. Galloway FJ, Denborough MA (1984) Phosphorus[31] nuclear magnetic resonance studies of muscle metabolism in malignant hyperpyrexia. Br J Anaesth 56:663–664
31. Radda GK, Taylor DJ (1985) Applications of nuclear magnetic resonance spectroscopy in pathology. Int Rev Exp Path 27:1–58
32. Kolb ME, Horne ML, Martz R (1982) Dantrolene in human malignant hyperthermia. Anesthesiology 56:254–262
33. Austin KL, Denborough MA (1977) Drug treatment of malignant hyperpyrexia. Anaesth Intensive Care 5:207–213

CHAPTER 13

Malignant Hyperthermia: Pre- and Post-Dantrolene
A Survey of the Greater Kansas City Area: 1965–1985

Mark G. Zukaitis, George P. Hoech, Jr.,
and John D. Robinson

Introduction

Malignant hyperthermia (MH) is a rare myogenic hypermetabolic syndrome that is associated with certain commonly used anesthetic agents and muscle relaxants. Although the first case of MH was noted over 60 years ago, the syndrome was not described in the literature until 1960 by Denborough and Lovell.[1] Since then, the pathophysiology and etiology have been studied and reviewed extensively. Treatment of MH before 1975 was primarily oriented toward correction of physiologic derangements. Mortality rates were characteristically 60% to 80%. Dantrolene sodium was added to the treatment protocol following successful use in MH-susceptible (MHS) pigs by Harrison[2] in 1975 and later that same year in MHS humans by Britt.[8] Since its approval by the FDA in 1979 for use in the treatment of MH, dantrolene has led to a dramatic decrease in morbidity and mortality caused by MH. This survey evaluates the pre–dantrolene and post–dantrolene experience with MH in the greater Kansas City area.

Methods

The anesthesiology departments of the hospitals in the greater Kansas City area were contacted and asked to submit reports on all MH cases that occurred between 1965 and 1985. In addition, information about the total number of anesthetics administered per year was requested. In years where exact data were not available, estimates of the number of anesthetics administered were requested. Case reports submitted were reviewed by the authors, and the anesthesiologists involved were interviewed. Following this in–depth review, 38 case reports were accepted for this survey.

TABLE 13.1. Characteristics of malignant hyperthermia patients

Sex	No. of patients	Age range	Median age
Female	18	2–32	5
Male	17	1–70	19

Several case reports were excluded based on insufficient information or absence of adequate diagnostic criteria for MH.

Results

All of the area anesthesiology departments participated in the survey. Thirty-five patients were identified as having MH based on thorough review of the case reports and interviews. A nearly even distribution of male (17) and female (18) patients was noted. Ages ranged from 1 to 70 years, with 80% less than 30 years old (Table 13.1). Of the 35 MH patients, 6 (18%) had a positive family history for MH (Table 13.2). One patient was adopted and had no family history available.

During the 20 year reporting period from 1965 to 1985, there were 38 MH episodes reported in these 35 patients. In this period, 2,038,200 anesthetics were administered for an incidence of 1:53,636. In addition, 12 anesthetics were administered to known MHS patients in which a MH episode was not triggered. Six additional patients received anesthesia who had no documented MH episode but who had a positive MH family history. The overall incidence of exposure to potential MHS patients was 1:36,396 (Table 13.3).

The anesthetic techniques used in the 38 MH episodes are shown in Table 13.4. Succinylcholine was used in 87% of the cases. A potent inhalation anesthetic (halothane, enflurane, isoflurane) was used in 84% of the cases. Halothane with succinylcholine was the most common technique, being used in 60% of the cases. Masseter spasm or generalized muscle rigidity was present in 30% and 21%, respectively, of the reported cases. The anesthetic technique and its relationship to muscle spasm or rigidity are seen in Table 13.5. Eighty percent of the patients with rigidity

TABLE 13.2. Incidence of a positive family history

Malignant hyperthermia patients	Positive family history	% with positive family history
35[a]	6	18

[a]One patient adopted.

TABLE 13.3. Malignant hyperthermia episodes related to general anesthetics administered in the greater Kansas City area (1965 to July 1985)

	General anesthetics	Incidence
	2,038,200	
MH episodes	38	1 per 53,636
MH episodes		
Repeat exposures	56	1 per 36,396
Family history of MH		

received halothane with succinylcholine. Twenty-seven patients (77%) had an MH episode on their first anesthetic exposure. The remaining seven patients (23%) had a total of 17 previous anesthetics without documented evidence of MH prior to their eventual triggering episode. Eight MH patients who triggered with a previous anesthetic received one or more repeat anesthetics (Table 13.6). On three occasions, a second MH episode occurred (Table 13.7). In two of these three cases, the patient received one or more triggering agents. In the third case, a narcotic, nondepolarizing muscle relaxant, nitrous oxide, oxygen technique was used following per os dantrolene pretreatment. In spite of this, the patient exhibited tachycardia, premature ventricular contractions, acidosis, and fever. The case was cancelled promptly. Several days later, the same patient was given per os dantrolene preoperatively, and a continuous intravenous infusion of dantrolene was given during induction with the same balanced anesthetic technique. The patient did not trigger. Sixteen patients with either a positive personal or family history received prophylactic dantrolene (Tables 13.8 and 13.9). Dantrolene prophylaxis was effective in preventing MH in all but the one case previously mentioned.

There were four deaths overall from MH during the years reviewed.

TABLE 13.4. Muscle relaxant and anesthetic agent use in 38 malignant hyperthermia episodes

Muscle relaxant	Anesthetic agent	No. of patients with MH
Succinylcholine	Halothane	23
Succinylcholine	Nitrous oxide	4
Succinylcholine	Enflurane	4
Succinylcholine	Isoflurane	2
Pancuronium	Ketamine and nitrous oxide	2
Pancuronium	Enflurane and halothane	1
Pancuronium	Enflurane	1
-	Halothane	1

TABLE 13.5. Masseter spasm and generalized muscle rigidity in 38 episodes of malignant hyperthermia

Muscle relaxant	Anesthetic agent	No. of patients with MH	Masseter spasm	Generalized muscle rigidity
Succinylcholine	Halothane	23	8	4
Succinylcholine	Nitrous oxide	4	2	1
Succinylcholine	Enflurane	4		
Succinylcholine	Isoflurane	2		
Pancuronium	Nitrous oxide	2		2
Pancuronium	Enflurane and Halothane	1		
Pancuronium	Enflurane	1		
	Halothane	1	1	1

The mortality rates for MH episodes in the pre-dantrolene and post-dantrolene periods were 31% and 0, respectively (Table 13.10). Overall, the mortality rate for the entire period was 11% (4/38). An analysis of the four deaths that occurred shows the diagnosis of MH was made late in the course of the episode, and appropriate therapy was delayed and ineffective. In one patient, the diagnosis of MH was made just prior to cardiac arrest, and in the other three, the diagnosis was made only after cardiac arrest had occurred. There were no survivors of MH when associated with cardiac arrest (Table 13.11).

TABLE 13.6. Previous exposure of eight malignant hyperthermia patients to muscle relaxants and anesthetic agents

	Previous exposure			Malignant hyperthermia exposure	
Patient	Muscle relaxant	Anesthetic agent	No. of times	Muscle relaxant	Anesthetic agent
1		Halothane	3	Succinylcholine	Halothane
2		Halothane and other general anesthetics	2 5	Succinylcholine	Ethrane
3		General anesthetic	1	Pancuronium	Halothane
4	Succinylcholine[a]	Halothane	1	Pancuronium	Ketamine and nitrous oxide
5		General anesthetic	1	Pancuronium	Enflurane
6		General anesthetic	1	Succinylcholine	Halothane
7		Halothane	3	Succinylcholine	Halothane
8		Halothane	1	Succinylcholine	Halothane

[a]In retrospect, this patient had triggered but had not been diagnosed as MH.

TABLE 13.7. Repeat anesthetic and muscle relaxant exposure in eight patients after a malignant hyperthermia episode

| | Malignant hyperthermia episode | | Repeat exposure | | | |
Patient	Muscle relaxant	Anesthetic agent	Muscle relaxant	Anesthetic agent	Dantrolene pretreatment	MH triggered
1		Halothane	Succinylcholine	Halothane	No	Yes
2	Succinylcholine	Halothane	Metrocurine iodide	Nitrous oxide	No	No
3	Pancuronium	Ketamine and nitrous oxide	A. Pancuronium	Nitrous oxide	Yes[a]	Yes
			B. Pancuronium	Nitrous oxide	Yes[a,b]	No
4	Succinylcholine	Halothane	Pancuronium	Nitrous oxide	Yes[b]	No
5	Succinylcholine	Halothane	Pancuronium	Nitrous oxide	Yes[b]	No
6	Succinylcholine	Nitrous oxide		Nitrous oxide	No	No
7	Succinylcholine	Nitrous oxide	Succinylcholine	Nitrous oxide	No	Yes
8	Pancuronium	Enflurane	Metrocurine iodide	Nitrous oxide	Yes[a,b]	No

[a]Orally.
[b]Intravenously.

TABLE 13.8. Indications for prophylactic use of dantrolene sodium in 16 patients

	Patient history	Family history
Diagnostic malignant hyperthermia	7	3
Suggestive for malignant hyperthermia	4	2

TABLE 13.9. Dantrolene sodium pretreatment in 16 patients at risk of developing malignant hyperthermia

Mode of administration	Dantrolene sodium pretreatment	Patients in whom MH was triggered
Oral	9	1
Intravenous	5	0
Oral and intravenous	2	0

TABLE 13.10. Malignant hyperthermia death rate in 38 episodes predantrolene and postdantrolene sodium

	Malignant hyperthermia episodes	Deaths	% Mortality
Predantrolene sodium 1965–1980	13	4	31
Postdantrolene sodium 1980–1985	25	0	0

Discussion

As more information about MH becomes available, it is clear that it is not a unique entity, but rather a complex syndrome with a wide spectrum of severity. In his work with MHS Poland China pigs, Williams[4] found a large variation in sensitivity to triggering agents. Some animals will spontaneously trigger in the absence of triggering agents. Others will survive without treatment merely by removal of the triggering agent early in the episode. This concept has been further substantiated in humans by Ording.[5] In his study of MH in Denmark, he found distinct groups of MH patients whom he referred to as "fulminant" versus "abortive" cases.

In our survey, we too have seen a wide range of susceptibility. Some patients develop tachycardia and muscle rigidity early, and biochemical derangements become quite pronounced. Others seem to smolder for hours before becoming critically ill. Some MH episodes were resolved merely by ending the anesthetic exposure. Yet, even avoidance of triggering agents and pretreatment with dantrolene appears not to have been 100% protective. Malignant hyperthermia in humans may have a spontaneous triggering mechanism. This has been suggested by Wingard[6] and Williams[7] invoking stress or release of endogenous catecholamines as the triggering event.

Because of this wide range of susceptibility, MH has made categorization

TABLE 13.11. Analysis of four deaths from malignant hyperthermia

	No. of patients
Succinylcholine administered	3
Halothane administered	4
Malignant hyperthermia suspected or diagnosed before cardiac arrest	1
Dantrolene sodium administered	0
Survivors of cardiac arrest	0

difficult. The incidence, age, and sex distribution found in this study compare favorably to other reviews by Britt et al.[8] and Ording.[5] There was a substantial increase in case reports in the last 5 years compared to the first 15 years. This is difficult to explain unless one considers the human aspects of reporting. Significant efforts have been made to educate and inform the anesthesia community about MH. Better informed practitioners make better diagnosticians. Greater awareness is the key. Electrocardiographic and temperature monitoring as well as arterial blood gas analyses are now the standard of care during anesthesia. These techniques assist the well-informed anesthetist in making an early diagnosis and treating this deadly syndrome early. It is our belief that some episodes of MH may have gone unrecognized, particularly in the pre-dantrolene period. The exact incidence of MH would be expected to be higher than that reported here.

Succinylcholine and potent inhalation anesthetics were found to be the major triggering agents. This is similar to findings reported elsewhere.[5,8] Avoidance of these drugs in the routine practice of anesthesiology should significantly decrease the incidence of MH. Dantrolene was used successfully to treat MH episodes. No deaths occurred in our survey in patients who received dantrolene. It represents the most significant breakthrough in the treatment of MH to date. Dantrolene pretreatment was found to be very effective also in preventing MH. Because of these factors, dantrolene should be readily available in all anesthetizing locations. In spite of dantrolene's effectiveness, education, awareness, and vigilance always remain the cornerstone in the treatment of MH.

References

1. Denborough MA, Lovell RR (1960) Anaesthetic deaths in a family. Lancet 2:45
2. Harrison GG (1975) Control of malignant hyperpyrexic syndrome in MHS swine by dantrolene sodium. Br J Anaesth 47:62–65
3. Britt BA (1979) Etiology and pathophysiology of malignant hyperthermia. Fed Proc 38:44–48
4. Williams CH (1977) The development of an animal model for the fulminant hyperthermia-stress syndrome, in Henschel EO (ed) *Malignant Hyperthermia: Current Concepts.* New York: Appleton-Century-Crofts, pp 117–140
5. Ording H (1985) Incidence of malignant hyperthermia in Denmark. Anesth Analg 64:700–704
6. Wingard DW (1977) Malignant hyperthermia-acute stress syndrome of man? in Henschel EO (ed) *Malignant Hyperthermia: Current Concepts.* New York: Appleton-Century-Crofts, pp 79–95
7. Williams CH (1976) Some observations on the etiology of fulminant hyperthermia-stress syndrome. Perspect Biol Med 20:120–130
8. Britt BA, Kalow W (1970) Malignant hyperthermia: A statistical review. Can Anaesth Soc J 17:293–315

The Role of the Sympathetic Nervous System in Patients Susceptible to Malignant Hyperthermia

J. Hilary Green, F. Richard Ellis, P. Jane Halsall, and Iain T. Campbell

Introduction

Clinical signs of sympathetic overactivity are usually observed in humans during a malignant hyperthermia (MH) crisis. Due to logistic difficulties in studying MH crises in humans, it was decided to investigate the integrity of the sympathetic nervous system during stress in humans susceptible to MH. Four major studies have now been carried out at the MH Investigation Unit in Leeds to assess the role of the sympathetic nervous system to physiologic stress. Three of these studies involved exercise, and the fourth involved surface cooling as the physiologic stressor.

Rodahl et al.[1] in a study of heavy work of short duration showed that a decrease in plasma free fatty acid (FFA) was followed by a rise during recovery, whereas a steady rise in FFA was seen during prolonged work of low intensity. In the same study, different patterns of lactate response were associated with the different exercise regimens; higher plasma lactate concentrations were seen with high-intensity work. The increase in plasma catecholamines was also related to work intensity, with higher levels in response to high-intensity work. Catecholamines are thought to play an important role in fuel mobilization.[2] Thus, the exercise studies were designed to vary the intensity and duration of work in order to examine the effect of different patterns of fuel mobilization in control and MH-susceptible (MHS) patients.

Methods

Subjects

Two groups of subjects took part in each study. One group was shown to be MH susceptible following an in vitro investigation of muscle samples according to an internationally agreed protocol.[3] The second control group, who had not been screened for MH, was assumed to be normal. The ages, heights, and weights were similar in the two groups for each study. The same subjects did not take part in every study. All those who took part gave written informed consent to the procedures, which had been approved by the Leeds Eastern Area Health Authority Ethics Committee.

Test Procedures

Exercise Study 1

The first exercise protocol involved 20 minutes of light exercise (67 W) in both the fasted and postprandial state.[4] Metabolic rate was calculated from measurements of respiratory gas exchange using Weir's formula.[5] Blood samples were withdrawn for the measurement of circulating glucose, lactate, pyruvate, triglycerides (TG), insulin, cortisol, thyroxine (T_4), and growth hormone (GH).

Exercise Study 2

The second exercise study involved periods of progressively severe exercise up to an intensity to give heart rates above 180 bpm.[6] Body temperature was measured at five sites (rectum, external auditory meatus, chest, thigh, and thumb). Blood samples were withdrawn for the measurement of circulating lactate, FFA and cortisol. A urine sample was taken on arrival in the laboratory and again 30 minutes after exercise for the measurement of epinephrine, norepinephrine, and dopamine.

Exercise Study 3

The third exercise study involved 120 minutes of level treadmill walking at 40% maximum oxygen consumption.[7] Maximum oxygen consumption was estimated by extrapolation of measurements of oxygen consumption (VO_2) and heart rate made during a preliminary exercise test comprising three workloads of increasing intensity. It was assumed that maximum heart rate is 220 bpm minus the subject's age in years. Body temperature was measured in the external auditory meatus. Oxygen consumption was calculated from measurements of respiratory gas exchange. Sweat evaporation rate was measured from the back and arm by dewpoint hygrometry.[8] Infrared radiation was measured from the back and arm using a handheld radiation thermometer. Blood samples were withdrawn for the measurement of plasma glucose, lactate, pyruvate, FFA, epinephrine, and norepinephrine.

Cooling Study

In the fourth study, the subjects were cooled using a liquid-conditioned coverall (RAF Institute Aviation Medicine) in which the liquid perfusing the coverall was lowered from approximately 25°C to 14°C.[9,10] Body temperature was measured at five sites: external auditory meatus, chest, arm, thigh, and calf. Heat production was calculated from measurements of respiratory gas exchange using Weir's formula.[5] Blood samples were taken for the measurement of circulating glucose, lactate, pyruvate, T_4, epinephrine, and norepinephrine.

Results

In the first exercise study, the dietary-induced thermogenesis was obtunded during exercise in MH patients. The insulin levels were significantly higher in the MH group after exercise in the fed state, and the cortisol levels were lower. The glucose/insulin ratio levels were significantly lower in the MH group at rest in the fasted and fed state and after exercise in the fasted state. None of the other variables were significantly different between the groups.

During a progressive exercise test, up to a theoretically maximal heart rate, a different pattern of peripheral body temperature change was seen in MH patients compared with control subjects. Plasma lactate, FFA, and cortisol concentrations were higher in the MHS group compared with the control group at various times during the study. However, as a whole, the study revealed no consistent differences between the two groups.

In the third exercise study, there were no differences between the MHS and normal patients in any of the variables measured.

In the cooling study, core temperature increased in both the MH and control subjects at the start of the cooling periods, with a greater increase occurring in the MH group. Although there were no significant differences between groups in the other variables measured, the increase in plasma norepinephrine of the MH group was greater than in most of the control subjects.

Discussion

In the first exercise study, the significantly lower glucose/insulin ratio (due to high insulin levels) would support the hypothesis that MHS patients display a different sympathetic response to exercise than do normal subjects. However, these data suggest a smaller sympathetic response in MHS people, since norepinephrine inhibits insulin release from the B cells of the pancreatic islets of Langerhans[11] by its action on alpha adrenoceptors.

The lower oxygen consumption in response to exercise in the fed state, compared with the control subjects, would be consistent with a greater sympathetic response in the control subjects.

In the second exercise study, the elevated lactate and FFA levels in MH compared with control subjects are consistent with increased plasma epinephrine.[12] Elevated levels of lactate and FFA are associated with a rise in metabolic rate,[12] although an elevated level of FFA is associated with an increased metabolic rate only with concomitant sympathetic stimulation.[13] A greater rise in metabolic rate in the MH group would account for the elevated core temperature that was observed if heat loss was either the same as or greater than that in the control group. However, it is interesting that urinary excretion of norepinephrine actually decreased after exercise in both groups when one might have expected an increase with an increase in sympathetic nervous activity.

In the first of these exercise studies, subjects completed the same work and therefore comparisons were made between subjects at the same workload. It is well-documented that the sympathoadrenal response to exercise is modified by endurance training.[14-17] Therefore, in the third exercise study, subjects exercised at the same workload relative to their exercise capacity. In this way, any differences in sympathetic response due to individual fitness were eliminated, since at the same relative workload plasma norepinephrine levels are unchanged by physical training.[18] In the third exercise study, 2 hours of treadmill walking produced no differences in any of the variables between control and MH subjects.

These exercise studies suggest that neither short duration-high intensity nor long duration-low intensity exercise need be contraindicated for MH patients. The possibility that exercise in the heat or under the more stressful conditions of competition might be harmful to MH patients has not been determined by this work.

In the fourth cooling study, core temperature increased more in the MH patients than in the control subjects. It could be speculated that this may be associated with a more intense cutaneous vasoconstriction, under the control of sympathetic noradrenergic nerves, compared with control subjects. Our measurements of peripheral blood flow were hampered due to shivering.[9] However, the increase in plasma norepinephrine of the MH group was greater than most of the control subjects.

Measurements of circulating norepinephrine concentration do not necessarily reflect norepinephrine turnover in the sympathetic nerve terminal[19] and measurements of circulating epinephrine do not necessarily reflect sympathetic nervous system activity.[20] However, there is a significant correlation between the rate of spontaneous nervous discharge along a sympathetic nerve in skeletal muscle and circulating norepinephrine.[21] These results, therefore, suggest a difference in sympathetic activity in response to cooling in MH patients compared with most of the control subjects. Further studies are needed to test this hypothesis further and to determine the source of the norepinephrine.

There is no confirmation of increased sympathoadrenal activity related either to exercise or to thermic stress in MH patients who were not undergoing an MH episode at the time. Rather, some of the observations in these studies are consistent with suppressed sympathetic nervous activity in MH patients during stress. This would not be unreasonable, since a variety of stresses (fasting, hypoglycemia, acute moderate hypoxia, vasovagal trauma, and acute trauma up to 12 hours after injury) are associated with depressed sympathetic nervous activity with a concomitant rise in adrenal medullary secretions of catecholamines.[20] The importance of this work could be that we have so far obtained no evidence of an awake triggering mechanism due to abnormal catecholamine release. This has wide implications in terms of the lifestyle of MH patients.

Summary

Four studies have been carried out to compare the thermoregulatory, hormonal, and metabolic responses to stress in patients susceptible to MH with those of normal individuals. Three of the stress tests involved exercise, and the fourth involved cooling. None of the studies revealed any consistent differences between MHS and control subjects. In particular, there was no clear evidence of an enhanced sympatho adrenal response to these stress tests in patients susceptible to MH.

References

1. Rodahl K, Miller HI, Issekutz B, Jr (1964) Plasma free fatty acids in exercise. J Appl Physiol 19:489–492
2. Christensen NJ, Galbo H (1983) Sympathetic nervous activity during exercise. Ann Rev Physiol 45:139–153
3. European Malignant Hyperpyrexia Group (1984) A protocol for the investigation of malignant hyperpyrexia (MH) susceptibility. Br J Anaesth 56:1267–1269
4. Campbell IT, Ellis FR, Evans RT (1981) Metabolic rate and blood hormone and metabolic levels of individuals susceptible to malignant hyperpyrexia and rest in response to food and mild exercise. Anesthesiology 55:46–52
5. Weir V De JB (1949) New methods for calculating metabolic rate with special reference to protein metabolism. J Physiol (Lond) 109:1–9
6. Campbell IT, Ellis FR, Evans RT, Mortimer MG (1983) Studies of body temperatures, blood lactate, cortisol and free fatty acid levels during exercise in human subjects susceptible to malignant hyperpyrexia. Acta Anaesth Scand 27:349–355

7. Ayling JH, Ellis FR, Halsall PJ, Campbell IT, Currie S Caddy J (1985) Thermoregulation and plasma catecholamines during submaximal work in individuals susceptible to malignant hyperpyrexia (Abstr). Med Sci Sport and Exercise 17:274

8. Lamke LO (1970) An instrument for estimating evaporation from small skin surfaces. J Plast Reconst Surg 4:1–7

9. Ayling JH, Currie S, Ellis FR, Hasall PJ, Hay E, Hills R (1983) The influence of surface cooling on body temperature and heat production in patients susceptible to malignant hyperpyrexia (Abstr). J Physiol (Lond) 348:34P

10. Ayling JH, Ellis FR, Halsall PJ, Currie S (1985) Thermoregulatory responses to cooling in patients susceptible to malignant hyperpyrexia. Br J Anaesth 57:873–990

11. Unger RH, Rouiller D, Orci L (1981) Control of fuel fluxes by polypeptides, in Bloom FE (ed) Peptides: Integrators of Cell and Tissue Function. New York: Raven, pp 69–79

12. Svedmyr N (1966) Studies on the mechanism for the calorigenic effect of adrenaline in man. Acta Physiol Scand 68:84–95

13. Kjekshus JK, Ellekjaer E, Rinde P (1980) The effect of free fatty acids on oxygen consumption in man: The free fatty acid hypothesis. Scand J Clin Lab Invest 40:63–70

14. Bloom SR, Johnson RH, Park DM, Rennie MJ, Sulaiman WR (1976) Differences in the metabolic and hormonal response to exercise between racing cyclists and untrained individuals. J Physiol (Lond) 258:1–18

15. Cousineau DR, Ferguson J, De Champlain J, Gauthier P, Cote P, Bourassa M (1974) Catecholamines in coronary sinus during exercise in man before and after training. J Appl Physiol: Resp Environ Exercise Physiol 43:801–806

16. Hartley LH, Mason JW, Hogan RP, Jones LG, Kotchen TA, Mougey EH, Wherry FE, Pennington LL, Ricketts PT (1972) Multiple hormonal responses to prolonged exercise in relation to physical training. J Appl Physiol 33:607–610

17. Winder WW, Hagberg JM, Hickson RC, Ehansi AA, McLane JA (1978) Time course of sympathoadrenal adaptation to endurance exercise training in man. J Appl Physiol: Resp Environ Exercise Physiol 45:370–371

18. Peronnet, F, Cleroux J, Perrault H, Cousineau D, Champlain JD, Nadeau R (1981) Plasma norepinephrine response to exercise before and after training in humans. J Appl Physiol: Resp Environ Exercise Physiol 51:812–815

19. Esler M, Jennings G, Korner P, Blombery P, Sacharias N, Leonard P (1984) Measurement of total and organ-specific norepinephrine kinetics in humans. Am J Physiol 247 (Endocrinol Metab 10):E21–E28

20. Young JB, Rosa RM, Landsberg L (1984) Dissociation of sympathetic nervous system and adrenal medullary responses. Am J Physiol 247:(Endocrinol Metab 10):E35–E40

21. Wallin BG, Sundlof G, Eriksson BM, Dominiak P, Grobecker H, Lindblad LE (1981) Plasma noradrenaline correlates to sympathetic muscle nerve activity in normotensive man. Acta Physiol Scand 111:69–73

Index

A

A23187 ionophore, 70, 131
Acetylpromazine, 110
Actin, 143
Alpha receptor, 2, 13
Amines, biogenic, 20–28
4-Aminopyridine (4-AP), 83
Anesthesia
  general, 62–71
  halothane, *see* Halothane
  sevoflurane, 50–56
Anesthetic related equine myopa-
    thy, 100–103
Antioxidant system deficit, 68–69
4-AP (4-aminopyridine), 83
Arachidonic acid, 68
Arterial hypotension, 101–102

B

Biochemistry
  of equine myopathy, 105–106
  of malignant hyperthermia, 142–
    143
Biogenic amines, 20–28
Biopsies, muscle, 128
Blood cells
  red, 134
  white, 135
Blood components, tests using,
    134–135
Blood pressure data, 31, 32
Butorphanol, 110

C

Caffeine, 65, 70
  halothene and, *see* Halothane-
    caffeine contracture test, Hal-
    othane-caffeine specific con-
    centration
Caffeine contracture test, 122, 129
Caffeine-specific concentration
    (CSC), 110, 129–130
Calcium channel activity, 16
Calcium content of MHS muscle,
    67
Calcium entry blockers, 88
Calcium ion, 2–3
  intracellular, 134
Calcium ion release, 61
  Endo's calcium-induced, 70
  halothane and, 65–66
Calcium ion uptake test, 132–133
Calmodulin, 88
Calsequestrin, 60–61
Carbamylcholine, 78–84
Cardiac dysrhythmias, 31
Cardiac index, 31, 36
Cardiac work indices, 31, 40
Cardiovascular responses, 93
Catecholamines, 88
  plasma, 7–17
Cell membrane functional charac-
    teristics, 2–3
Central-core disease, 142
Chloralhydrate, 110
CK, *see* Creatine kinase

Contracture tests, 121
    considerations, 123–127
    halothane-caffeine, *see* Halo-
        thane-caffeine contracture
        test
    important factors in, 131–132
    methodology, 128–132
    in stages of development, 130–
        131
Cooling study, 157, 158
Core temperature, development
    of, 33, 42–43
CPK (creatinephosphokinase), 141
Creatine kinase (CK)
    resting, 111, 135
    serum, 118
Creatinephosphokinase (CPK),
    141
CSC (caffeine-specific concentra-
    tion), 110, 129–130
Curare, 132

D
Dantrolene, 63, 67, 90–91, 143,
    144
    horses and, 104–105
    humans and, 147–153
    indications for use of, 151
    intravenous, 104–105
    pretreatment with, 151
Death rate in malignant hyperther-
    mia, 152
Deiodinases, 56
DHBA (3,4-dihydroxybenzylam-
    ine), 21
Diagnosis
    laboratory methods for, 121–136
    MH equivocal (MHE), 125
Diagnostic test development, 121–
    123
Diethyl ether, 63, 68
3,4-Dihydroxybenzylamine
    (DHBA), 21
Diltiazem, 40–41, 86–88

Dog, malignant hyperthermia in,
    118–119
Dynamic halothane contracture
    tests, 122, 129
Dysrhythmias, cardiac, 31

E
ECD (electrochemical detection),
    20–28
Effective dose (ED), 82
Electrochemical detection (ECD),
    20–28
Electromyography, 133
Endo's calcium-induced calcium
    release, 70
Endurance training, 158
Enflurane, 63
EOF (erythrocyte osmotic fragili-
    ty), 119
Epinephrine, 2
    in acidic phase, 20–28
    circulating, 158
    in pig plasma, 7–17
Equine malignant hyperthermia
    halothane-succinylcholine anes-
        thetic challenge in, 113
    histochemical analysis in, 112–
        113
    muscle biopsy studies in, 106–
        112
Equine myopathy, 100–114
    anesthetic related, 100–103
    biochemical studies of, 105–106
    exercise-related, 103–104
    MH and, 103
Erythrocyte osmotic fragility
    (EOF), 119
Excitation-contraction coupling,
    61, 144
Exercise myopathy, horse, 91–96
Exercise-related equine myopa-
    thy, 103–104
Exertional myopathy, 103–
    104

F
Fatty acids
free (FFA), 155–158
long-chain, 68
Free fatty acids (FFA), 155–158
Free sarcoplasmic reticulum, 60

G
Glucose/insulin ratio, 157–158
Glycoprotein, 53,000-dalton, 60–
61

H
Halothane, 62–63
calcium ion release and, 65–66
carbamylcholine and, 78–84
concentration, 132
equine myopathy and, 100–114
horse studies with, 90–97
spectrum of response to, 123–
125
succinylcholine and, 107, 113,
122, 130–131, 148–152
Halothane-caffeine contracture
test, 101
of equine muscle, 106–112
Halothane-caffeine specific concentration (HCSC), 122, 130
Halothane contracture tests
dynamic, 122, 129
static, 122–123, 128–129
Halothane-succinylcholine anesthetic challenge, 113
HCSC (halothane-caffeine specific
concentration), 122, 130
Heart rate, 31, 38
Heavy sarcoplasmic reticulum
(HSR), 66
Hemidiaphragms, rat, 80, 82–84
Hemodynamics in pigs, 30–44
High performance liquid chroma-
tography (HPLC), 20–28
Histochemical analysis in equine

malignant hyperthermia, 112–
113
Horse(s)
as animal models for MH, 100–
114
dantrolene and, 104–105
exercise myopathy, 91–96
in studies of MH, 90–97
HPLC (high performance liquid
chromatography), 20–28
HSR (heavy sarcoplasmic reticulum), 66
Humans, dantrolene and, 147–
153
Hypercapnia, 113
Hyperpyrexia, 141; *see also* Malignant hyperthermia
Hyperthermia, malignant, *see* Malignant hyperthermia
Hyperthyroidism, 55
Hypotension, 101–102

I
Inner ring deiodinase, 56
Inositol triphosphate, 3
Insulin and glucose, ratio of, 157–
158

K
K phenotype, 127
Kansas City survey, 147–153

L
Laboratory methods for diagnosis,
121–136
Light sarcoplasmic reticulum
(LSR), 66
Linoleic acid, 68
Long-chain fatty acids, 68
LSR (light sarcoplasmic reticu-
lum), 66
Lymphocytes, 135

M

Magnesium content of MHS muscle, 67

Malignant hyperthermia (MH)
  in animals, 142
  biochemistry of, 142–143
  characteristics of patients, 148
  clinical presentation of, 142
  death rate in, 152
  in dog, 118–119
  equine, *see* Equine malignant hyperthermia
  equine myopathy and, 103
  heat production in susceptible muscle, 1–4
  hemodynamics in, 30–44
  horse in studies relative to, 90–97
  horses and ponies as animal models for, 100–114
  humans and, 147–153
  identification of susceptibility to, 143–144
  laboratory methods for diagnosis, 121–136
  myopathies that predispose to, 141–142
  in pig, *see* Pig
  rapid development of, 34
  review, 141–145
  sarcoplasmic reticulum membrane and, 59–72
  site of muscle cell abnormality in, 143
  skeletal muscle and, 141–142
  sympathetic overactivity and, 155–159
  thyroid hormones during, 46–57
  treatment of, 144
Malignant hyperthermia susceptible (MHS) pigs, 30–31
Masseter muscle rigidity (MMR), 130, 131, 148–150
Mean corpuscular fragility (MCF), 134

MH, *see* Malignant hyperthermia
MH – and MH + phenotypes, 123–127
MH equivocal (MHE) diagnosis, 125
MHS (malignant hyperthermia susceptible) pigs, 30–31
MHX pigs, 64
MMR (masseter muscle rigidity), 130, 131, 148–150
Monodeiodinases, 56
Muscle biopsies, 128
Muscle biopsy studies in equine malignant hyperthermia, 106–112
Muscle cell, striated, 59
Muscle cell abnormality, site of, 143
Muscle contracture tests, *see* Contracture tests
Muscle, drugs affecting, 78–89
Muscle fiber types, 112
Muscle injury, 101
Muscle ischemia, 101–102
Muscle plasmalemma, 67–68
Muscle proteins, abnormal skeletal, 133
Muscle relaxants, 149–151
Muscle tests, 123–134
Myoglobin, urine, 106
Myoglobinuria, 101
Myopathy, exercise, horse, 91–96
Myosin, 143

N

Nitrendipine, 61–62
Nitrous oxide, 149
NMR (nuclear magnetic resonance) spectroscopy, 133–134, 144
Norepinephrine, 2, 3, 88
  in acidic phase, 20–28
  circulating, 158–159
  in pig plasma, 7–17

Nuclear magnetic resonance
    (NMR) spectroscopy, 133–
    134, 144
Nucleotide depletion test, 134

O
Oleic acid, 68
Outer ring deiodinase, 56
Oxygen consumption, 31, 35, 156,
    158

P
Pale, soft, exudative pork (PSEP)
    syndrome, 142
Pancuronium, 149–151
PCr (phosphocreatine), 144
Peripheral resistance, total, 31, 39
Phenotypes
    K, 127
    MH − and MH + , 123–127
Phentolamine, 4
Phenylephrine, 4
Phosphatidyl inositol phosphate,
    2–3
Phosphofructokinase-fructose bi-
    phosphate cycle, 55
Phosphorylase, 143
Physical training, 158
Pig
    biogenic amines in, 20–28
    drugs affecting muscle of, 78–
        89
    heat production in susceptible
        muscle, 1–4
    hemodynamics in, 30–44
    malignant hyperthermia suscep-
        tible (MHS), 30–31
    MHX, 64
    plasma catecholamines in, 7–17
    thyroid hormones in, 46–57
Plasmalemma, muscle, 67–68
Platelet model systems, 134
Ponies, *see* Horse(s)

Porcine stress syndrome (PSS), 1,
    16, 142
Potassium pyroantimonate, 64
PSE (pale, soft, exudative) syn-
    drome, 142
PSS (porcine stress syndrome), 1,
    16, 142
Pulmonary vascular resistance,
    33, 36, 41
Pyroantimonate method, 64

R
Radioimmunoassay (RIA), 46
Rat, hemidiaphragms of, 80, 82–
    84
Red blood cells, 134
Relative standard deviation
    (RSD), 23
Relaxants, muscle, 149–151
Resting CK, 111, 135
Reverse triiodothyronine (rT₃),
    46–56
Reversed phase high performance
    liquid chromatography (RP-
    HPLC), 20–28
RIA (radioimmunoassay), 46
RP-HPLC (reversed phase high
    performance liquid chroma-
    tography), 20–28
RSD (relative standard deviation),
    23
rT₃ (reverse triiodothyronine), 46–
    56
Ruthenium red, 63–64

S
Sarcolemma, 59
Sarcoplasmic reticulum, 59–72
    free, 60
    heavy (HSR), 66
    light (LSR), 66
Sarcoplasmic reticulum mem-
    brane, 59–72

Sarcoplasmic reticulum vesicles, 60

SDS (sodium dodecyl sulfate) gel electrophoresis, 133

Sevoflurane anesthesia, 50–56

Skeletal muscle, 141–142

Skeletal muscle proteins, abnormal, 133

Skinned fiber test, 133

Sodium dodecyl sulfate (SDS) gel electrophoresis, 133

Static halothane contracture tests, 122–123, 128–129

Striated muscle cell, 59

Stroke volume index, 31, 37

Stroke work index, 31, 40

Succinylcholine, 1, 2, 78, 100
halothane and, 107, 113, 122, 130–131, 148–152

Susceptibility, identification of, 143–144

Sympathetic overactivity, 155–159

T

T tubules, 59

T_3 (triiodothyronine), 46–56

T_4 (thyroxine), 46–56

Tachycardia, 9–10, 16, 28, 35–36

Temperature, core, development of, 33, 42–43

Tetracaine, 63–64

Thymol, 68

Thyroid hormones, 55
plasma, during MH, 46–57

Thyroxine (T_4), 46–56

Transverse tubules, 59

Triads, 60

Trihydroxyindole procedure, 7

Triiodothyronine (T_3), 46–56
reverse (rT_3), 46–56

Troponin, 143

Troponin C, 65

U

Urine myoglobin, 106

V

Ventilation responses, 94

Verapamil, 96

W

White blood cells, 135

X

Xylazine, 110